THE NEXT TIME WE STRIKE

THE NEXT TIME WE STRIKE

LABOR IN UTAH'S COAL FIELDS, 1900-1933

ALLAN KENT POWELL

UTAH STATE UNIVERSITY PRESS
LOGAN UTAH

Second Printing, 1992

Library of Congress Cataloging in Publication Data

Powell, Allan Kent.
The next time we strike.

Includes index.
1. Trade-unions—Coal-miners—Utah—History.
2. Collective bargaining—Coal Mining industry—Utah—History. I. Title
HD6515.M616U87 1985 331.88'122334'09792 84-27005
ISBN 978-0-87421-161-0 pbk
ISBN 978-0-87421-934-0 e-book

For my parents, Leland and Luella Grange Powell, who met the dangers and injustices of coal mining with the conviction that education held the promise of a better life for their children.

Contents

Illustrations

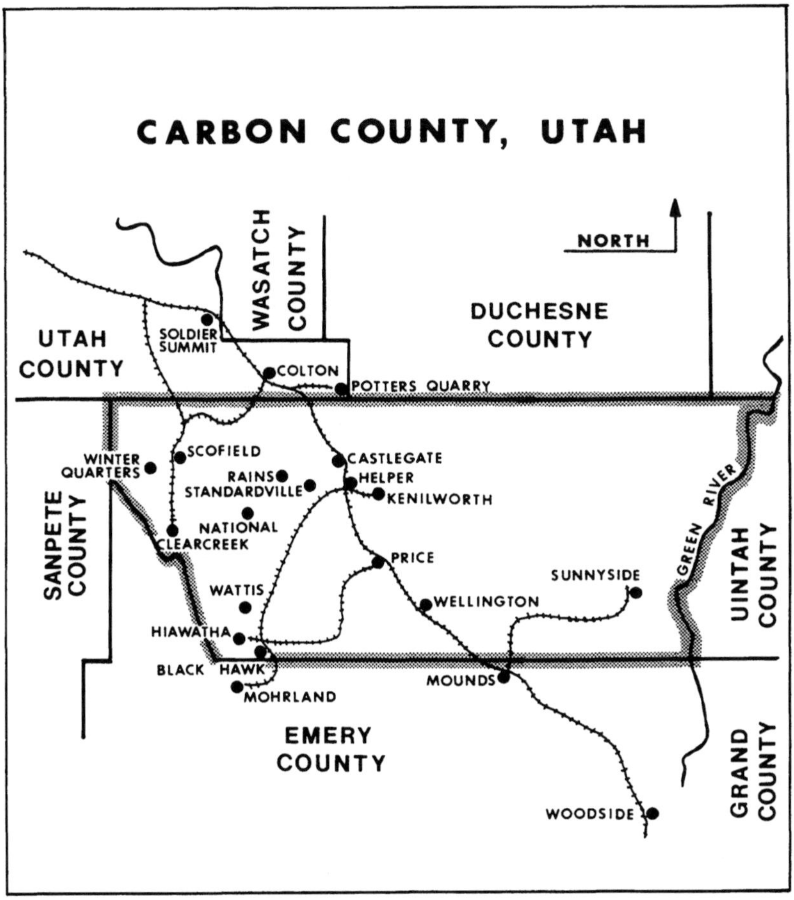

Map of Carbon County, Utah.

Acknowledgements

I am indebted to many friends and colleagues for their assistance in the preparation of this book. Philip F. Notarianni, Helen Zeese Papanikolas, and Charles S. Peterson have given unfailing support since I first began my study of labor in the Utah coal fields nearly a decade and a half ago. Each has made available valuable research notes and sources, read and provided substantive comments and suggestions on many drafts, and encouraged the publication of the book in every way possible. For their insights and help I am grateful, but recognize my sole responsibility for any errors that have been incorporated within the book.

Craig Fuller and Nancy Taniguchi have also shared important sources. Librarians and staff of the Utah State Historical Society Library, Utah State Archives, Historical Department of the Church of Jesus Christ of Latter-day Saints, University of Utah Library, Brigham Young University Library, University of Colorado Library, College of Eastern Utah Library, the National Archives, and the National Headquarters of the United Mine Workers of America all facilitated my research.

Four University of Utah professors, F. Alan Coombs, S. Lyman Tyler, Reed C. Richardson, and the late David E. Miller gave support, encouragement, and direction in the early days of the study. Colleagues in the Utah Labor History Forum, John R. Sillito, John S. McCormick, J. Kenneth Davies, Gregory Thompson, Jay M. Haymond, and Philip F. Notarianni have reassured me of the need for such a book and the value of studying Utah's labor history.

I appreciate the excellent work of the Utah State University Press staff, especially Linda Speth and Alexa West. They have gone far beyond the regular duties of a publisher in their quest to eliminate errors and inconsistencies and to improve the readability of the text.

Melvin T. Smith, Director of the Utah State Historical Society, through his commitment to professionalism, independent thought, personal initiative, and service to the people of Utah has helped me find my work in the Utah Historic Preservation Office very stimulating and rewarding. My always-supportive wife Brenda helped me find the necessary time for research and writing. She has done more than her share in managing our family affairs and wrangling our three children Lee, Liesel, and Adrianna.

Introduction

Over fifty years ago, in 1933, Utah coal miners were organized by the United Mine Workers of America (UMWA), and an agreement was negotiated with coal operators marking the first official recognition of the right of Utah miners to organize. Residents of Utah's coal region and others interested in Utah history have assumed that no union existed in Utah until President Franklin D. Roosevelt allowed John L. Lewis to send UMWA organizers to the state. However, the recognition of the Utah coal miners was not the spontaneous result of the New Deal. Rather, it was a reaction to the deaths of 200 miners in the Winter Quarters mine explosion in 1900, a series of strikes, as well as the culmination of more than fifty years of organizing attempts by the Knights of Labor, the United Mine Workers of America, the Western Federation of Miners, the Industrial Workers of the World, and the National Miners Union.

Utah's coal miners are aware of their history, but for the most part, it has been preserved and passed on in oral rather than written form. Their songs tell of past strikes and mine disasters. Their stories recount the abuses of coal companies, the dangers of coal mining, the financial difficulties during the lean years and slack seasons, and the intensity of labor conflicts.

Before the miners who worked and lived in Utah's coal fields pass on, it is important to capture their story in written form. The miners' oral traditions as well as a variety of coal company reports, labor records, government documents, newspaper articles, and secondary sources highlight several important themes. The most important from 1900 to 1933 include the following: the dangerous working conditions and the deeply felt tragedies from the disasters at Winter Quarters in 1900 and at Castle Gate in

1924; the generally conservative nature of miner demands during strikes; the commitment of Utah coal operators to keep Utah a nonunion state; the relatively nonviolent nature of Utah's labor confrontations in comparison with Colorado's and other coal region's; the intense belief in unionism by several people who dedicated their lives to the union struggle; the importance of the foreign-born miners in the fight for unionism; the significance of Utah in the nationwide organizing strategy of the United Mine Workers of America; and the conservative attitudes of leaders of the Church of Jesus Christ of Latter-day Saints (LDS or Mormon church) toward the union movement.

Above all, the United Mine Workers of America and a slow-growing union consciousness were positive forces for Utah and its coal miners. The union promoted greater class consciousness and achieved better conditions for the miners; however, it was conservative rather than revolutionary in its goals and practices. The union worked within the system for the betterment of its members by exerting collective pressure for more equitable and humane treatment. Although the union occasionally pressed for higher wages, most of the time it simply defended miners from proposed wage cuts or protected them against company abuses. In the 1930s, when the dire economic situation seemed to threaten the American system, the union helped prevent violence by urging members to demonstrate patience, to believe in the government, and to avoid violent and radical alternatives. In Utah's Carbon County, if the United Mine Workers of America had not offered a more conservative alternative to the revolution-oriented National Miners Union, civil war may have flared and perhaps spread to other parts of the state.

The events, spread over three decades, that culminated in the organization of Utah's coal miners were evolutionary rather than revolutionary. Agitated by a very real sense of fear after the Winter Quarters disaster in 1900, miners recognized their second-class status in the face of low pay, dangerous working conditions, and mistreatment. Striking first as a band of fellow workers, Utah miners had little hope for victory except through national affiliation and recognition of their union by the coal companies. Time and time again, they turned to the United Mine Workers of America for support. In 1901, Utah miners received no help; and in 1904, after a promising beginning, union support ended leaving a bitter and hostile legacy for the UMWA in Utah. However, in 1922 the union redeemed itself and, though not winning recognition, it kept a spark of goodwill alive in the hearts of most Utah miners. In 1933, when the choice was between the United Mine Workers of America and the National Miners Union, the UMWA won easily.

Ultimately, the UMWA helped reduce the animosity and antagonism among the ethnic groups in the coal camps. At first the work force was divided into two groups, the native-born miners and immigrants from English-speaking countries who became citizens in a short time and the non-English-speaking miners.

From American miners, who typically disdained the "spaghetti-eating, garlic-breathed foreigners" and who generally felt them unworthy of American citizenship, the union demanded cooperation with and equality for their co-workers. In addition, by helping the immigrant move into the mainstream of American life, the union eased nativistic suspicion.

For many members, the union was a school for learning democratic practices. Union elections taught nomination procedures, campaign techniques, and voting processes. Some, like the Italian-born Frank Bonacci, would move from union campaigns to state elections. In the union's demands for fair and equal treatment, all could observe the democratic principles of the American constitution and government.

Today, Utah's Carbon County has been praised as a model of cooperation and mutual respect among its different ethnic and religious groups. However, a few generations ago nativistic chauvinism and southeastern European pride formed a volatile and unstable compound that sporadically erupted in violence—often in connection with labor disputes. The tranquility that rules today is in no small measure based on the common experience forged in the UMWA.

The story of the Utah coal fields, first, reveals that the establishment of the union was a long and difficult process; second, illustrates the nature of labor relations in Utah as the state moved from the pioneer philosophy of a local, self-sufficient economy to one integrated with national and world economies; and finally, shows that the struggle was part of a larger saga as the development of the coal industry hastened the end of one of America's last frontiers. While Utah coal miners tried to organize and then achieve recognition of their union, cowboys carried on their traditional trade within sight of the mines; Indians executed their last violent resistance to the white man's intrusion on their land; and settlers filled up the last uninhabited spaces within the state. In a sense, the coal miners' fight was a response to the challenges of a new day, yet a frontier character permeates their story. The union victory in 1933 signified the emergence of Utah from a pioneer and frontier state to one marked by industrialization and urbanization. For Utah, that transition began as its coal fields were developed in the 1880s, as political decisions charted a new course in Mormon–non-Mormon relations, and as the last unoccupied area of the territory was settled.

The study of the organization of Utah's coal miners is important for several reasons. Although much has been written about individual coal strikes and national union leaders, there has been little analysis of the process of organization in a state—especially one for which union recognition came so late. Utah, therefore, is an excellent case study for examining and comparing the unionization process with other coal-mining areas. It is a study of a naive, idealistic, and ungainly movement that developed a focus, a cadre of leaders, and a consciousness that finally matured with the coming of the New Deal labor policies. Special attention is given to the attitudes of and support by national and district union leaders toward a nonunion region. The study explains why unionization was slow in coming to Utah—an analysis that might be helpful in explaining the success or failure of other organizational struggles in the state.

In considering the larger history of the United Mine Workers of America, it is essential to understand how local miners seeking affiliation perceived the national union. Such an insight can only be gained by examining a specific locale where the issues, the policies, and the extent of national support can be carefully analyzed. This study demonstrates the impact of national policies on an unorganized region and investigates how well the nonunion miners fared without the union in their struggles for recognition.

Although coal was a significant industry in Washington, Montana, Wyoming, Colorado, New Mexico, and Utah, it, as well as other labor-related issues, has generally been neglected by western historians. Although a few excellent studies have been published on labor in the metal mines and the tragic 1913–14 strike in Colorado, no volumes exist on labor and labor-management relations in the coal industry. This is unfortunate because each union experience in the coal-mining districts of the West was unique, yet intertwined with the experience of neighboring states. Hopefully, studies about the western coal industries and their interrelationships will be forthcoming. Undoubtedly, they would provide a more complete picture of the American West.

On the local level, labor history in Utah suffers from the same neglect, although for different reasons. Until recently, Utah historians have concentrated on Mormon, agrarian, and nineteenth-century themes. Except for a few articles in the *Utah Historical Quarterly* and one book, J. Kenneth Davies's *Deseret's Sons of Toil*, which deals with labor in Utah from 1852 to 1896, nothing has been published on Utah labor history. This study, by focusing on twentieth-century and non-Mormon themes, seeks a somewhat different vision of Utah's past and recognizes that the study of Utah's coal miners is essential to understanding Utah's development. As more scholars direct their studies to twentieth-century and non-Mormon topics, we may be able to understand more fully and document the state's complex development and heritage.

The Next Time We Strike is dedicated to many: to the descendants of those who worked the coal mines; to today's coal miners whose wages, benefits and improved working conditions were forged in the labor struggles of an earlier era; and to those who seek an understanding of the course of labor relations in one of Utah's and America's key industries.

Chapter I

A Legacy of Labor and Coal

Union-organizing efforts in the Utah coal fields took place in a state well known for its anti-union attitudes and where class tension and labor discord existed. Although the state's economy was founded on agriculture, commerce as well as mines, railroads, and developing cities constituted its future. Utah's leaders recognized that agriculture provided only a subsistence life while mining and manufacturing offered a prosperity that farming and ranching could not equal. As the long, hard-fought battles over polygamy and over church political control receded, a wave of aggressive capitalism swept over the state in the 1890s as Mormons and non-Mormons closed ranks to pursue economic gain and a higher standard of living—which often did not extend to the state's working class. The period was also characterized by a new generation of Mormon church leaders who, in practice though not in theology, rejected Joseph Smith and Brigham Young's communitarianism, and these new leaders had an important impact on the state's economic life.

Because most of the new church leaders came from the ranks of business, they held an inherent anti-union bias and feared that labor unions would destroy the loyalty of Mormon workers. Thus the Utah labor picture was a double exposure of the national scene—expanding industry, new railroads, immigrant labor, industrial towns, growing commercial centers, class tensions, and labor discord, over which was superimposed the local Mormon philosophy. This multifaceted image was characterized by the convictions of a chosen people: a belief that the righteous would prosper, an emphasis on obedience to church authority, and the elevation of the doctrine of individual free agency and responsibility over collective action and class consciousness. Labor unions, Mormon leaders held, prevented individual choice and furthermore were usually directed by men of questionable motives and character.

Yet contradictory elements in the Mormon heritage made the picture more complex and posed certain philosophical problems. The early development of the Church of Jesus Christ of Latter-day Saints was characterized by many of the same motives and objectives that nurtured the emerging American labor movement. A similarity between the church and the labor unions is suggested in the success of Mormon missionaries working among Europe's lower classes. The Mormon promise of temporal as well as eternal salvation sounded harmonious among the noise of trade unionists, socialists, communists, reformers, and others who addressed the plight of Europe's poor. Yet in America, Mormonism and the labor union movement seldom interacted because the LDS church was largely a frontier institution, and labor unions flourished in eastern cities and industrial pockets. Nevertheless, Mormon converts did not totally abandon their union ties; and for a time found tolerance, if not sympathy, among early leaders of the LDS church. In time, the objectives of organized labor were at odds with the new Mormon philosophical and economic outlook that stressed the church could provide all that was needed for its members. Labor unions, fraternal organizations, and other "secret societies" were merely distractions and hindrances for the Saints. But the reality of Utah's economic development, over much of which the Mormon church had little or no control, would find Utah's workers reaching out to unions.

In this milieu, the labor movement in Utah struggled along as part of the larger national story, attaining occasional victories, which encouraged continued efforts. At the same time, local issues and events forged a new mold for the Utah unionist. The nonunion coal miners indirectly benefited from the success of other workers and in times of open conflict were strengthened and supported by organized labor in Utah. When recognition came for the Utah coal miners in 1933, they moved easily into the mainstream of organized labor within the state.

The Labor Movement

The printers were the first group to organize in Utah. They established an informal association whose primary purpose was more social than economic.[1] The Deseret Typographical Union was organized on August 3, 1868, in Salt Lake City; and two weeks later, it received a charter from the National Typographical Union, one of the nation's oldest and most stable national organizations. Until 1880, the printers' union remained the only formally organized trade union in the Utah Territory. However, during the 1860s and 1870s, various crafts and trades established informal associations much as the printers had. On one occasion in 1864, one of these groups, the mechanics, attempted to expand into the economic realm by organizing a strike for higher wages.[2]

Until the completion of the transcontinental railroad in 1869, Utah's economy remained almost exclusively agrarian, despite church-directed attempts at organizing industries to produce goods for home consumption. Utah's mining era really began in 1863 when soldiers, under the command of Colonel Patrick Connor, discovered ore-bearing rock; but the inadequate transportation system precluded any significant mining development until after the coming of the railroad.

Mormon leaders responded to the arrival of the transcontinental railroad and to the corresponding threats of non-Mormon domination of the economy by organizing cooperative enterprises. During the cooperative movement in 1868, an estimated 200 separate enterprises ranging from cooperative mercantile stores to ironworks and textile factories were established within the sphere of Mormon colonization. The most interesting feature of the cooperative movement was the development of the United Order system which, in its purest form, obligated its members to donate all their property to the community, share equally in the communal produce, and live as one family.[3]

However, the church had directed economic affairs in the Utah Territory long before the beginning of the cooperative movement. The labor force, in particular, was subject to the direction of church leaders from the Mormon beginnings in the West until the end of colonization in the late 1870s. Skilled craftsmen were in demand to help strengthen the church; Brigham Young encouraged his missionaries to seek out the skilled craftsmen in the eastern states and in Europe. Once in Utah, these men were subject to be "called" by the church to settle in communities where their abilities and skills were most needed. For the unskilled and unemployed members, Mormon leaders provided public works projects, thus allowing everyone to earn sustenance.

Three factors explain the general lack of union organization before 1880: an agrarian economy with limited industrial development under the direction of the Mormon church; a Mormon antipathy to unions; and the lack of a large mining industry to encourage union organization among the miners. Between 1870 and 1880, Utah's economy changed significantly. Although by 1880 agriculture still employed the largest number of persons, 14,550, the number of miners rose dramatically from 519 in 1870 to 3,592 by 1880. Similarly, the number employed in the manufacturing and mechanical industries rose from 3,588 in 1870 to 7,630 in 1880.[4]

The first major labor trouble in Utah occurred in February 1881 in the Silver Reef Mining District of southwestern Utah. When stockholders exerted pressure because of reduced dividends, the Stormont Mining Company announced a 12.5 percent pay reduction for its workers. Some 250 miners struck and organized a local union, which exercised strong control for a short time before the union officers were arrested by a United States marshal in March 1881.

At the same time the Utah metal miners were attempting to organize, the state's railroad workers organized several different unions including the Brotherhood of Locomotive Firemen in 1882, the Order of Railway Conductors in 1883, and a Trainmen's Lodge in 1885. Through these railroad organizations the Knights of Labor made its way to Utah. Founded as a secret order in 1869 in Philadelphia at a tailors' meeting called by Uriah S. Stephens, the Knights of Labor was reorganized a decade later, under the leadership of Terence V. Powderly as a national industrial union open to all workers. With an effective central leadership, the union grew at a phenomenal rate. As the first national organization open to all workers, it preceded by seventeen years the American Federation of Labor organized in 1886.

The Knights of Labor was generally idealistic and humanitarian in its approach to organizing and directing working-class power through cooperative effort. With this philosophy, the Knights discouraged strikes, which were viewed, at best, as efforts to secure temporary relief. However, the arrival of the Knights of Labor in Utah coincided with the first recorded labor dispute in the eastern Utah coal fields. A strike during the winter of 1883 at the Pleasant Valley coal mines ended when Stake President Abraham O. Smoot journeyed from his Provo home to persuade the miners to return to work. Those who did not respond to this admonition were arrested by Sheriff J.W. Turner.[5]

The Knights of Labor presented the first serious effort of organized labor to arouse class consciousness and united action in the Utah Territory. The Knights was clearly an industrial union, unlike Utah's small craft unions, still in infancy during the 1880s. The organization also promoted a cooperative movement of its own, and until 1881 required its prospective members to pass through a secret initiation ritual. The principal aim of the Knights of Labor in Utah during the mid–1880s was to exclude Chinese workers from the railroads and mines in order to secure and preserve jobs for its members.[6] The anti-Chinese sentiment became so intense that in early September 1885, miners at Rock Springs, Wyoming, burned the community's Chinatown and murdered several Chinese. In the Utah coal fields, the Chinese were loaded in a boxcar and sent down the canyon.

The church opposed the violent methods used by the Knights to eliminate the Chinese and the radical terms in which some union representatives called for the overthrow of capitalistic monopolies. In addition, Mormon leaders charged that the Knights' secrecy and cooperative ideals threatened Mormon loyalty and cohesion. Spokesmen for the Mormon church concluded that the Knights of Labor and other socialist, anarchist, and internationalist groups were the organizations spoken of in the Book of Mormon established to achieve power through secret combinations. Such organizations, the Mormon scripture prophesied, would be destroyed by God.[7]

In the *Deseret News*, a church-owned newspaper, an editorial reiterated, "Any Mormon who would now apply for admission into the Order [Knights of Labor] or one who may be connected with it now and remain in it . . . would be left very much to himself."[8] However, by September 1886, when the editorial was published, the days of the Knights of Labor were numbered. The Chicago Haymarket Riot of May 4, 1886, was blamed on the Knights, and the organization quickly lost favor with the public. It also suffered major setbacks during the 1886 strikes in the Chicago stockyards and on the Gould railroad system in the Southwest. The cooperative enterprises proved unsuccessful, and the organization suffered from a widening breach between its leaders and members. Membership dropped from 700,000 in 1886 to 200,000 in 1888 and to 75,000 in 1893.[9] Despite its short life in Utah, the Knights of Labor played an important role in shaping the future of labor sentiment in the state. The ideas, objectives, and methods of the Knights of Labor threatened Mormon leaders to the point where church members were forbidden to take part in the activities of the labor organization. Consequently, the Knights left a legacy of mistrust for organized labor in Utah.[10]

Although the 1880s witnessed tension between the Mormon church and the industrially oriented Knights of Labor, craft unions increased in strength and number. In 1888, Salt Lake City witnessed its first Labor Day parade, and six months later, a Utah Federated Trades and Labor

Council was organized.[11] Also in 1889, members of the Stonecutters and Bricklayers Union employed by the Watson Brothers on the construction of the Zion's Savings Bank struck to secure the dismissal of all nonunion workers on the job. The strike was the first recorded attempt in Utah to establish a closed shop through a walkout.[12] The church mouthpiece, the *Deseret News*, saw great danger in the attempt to "dictate" to the contractors and branded it "tyranny in one of its worst forms . . . [which] strikes at the very foundation of human liberty. It creates a power that should not be exercised over the souls of men." The closed shop "would force men against their free choice to join in a society with which they are not in accord, and compel its members to a course . . . revolting to their sense of justice. . . ."[13]

The Zion's Savings Bank strike foreshadowed a general carpenters' strike in 1890 during which the boundaries of union jurisdiction in Utah were more clearly defined. The strike, which began on May 1, was part of the nationwide campaign for the eight-hour day. Although the contractors admitted the demands for increased wages and shorter hours were reasonable, they again balked at the carpenters' efforts to secure a closed shop and recognition of their union. The strike ended in a compromise, most contractors agreeing to deal with union men; however, the union was forced to recognize the employer's right to dismiss anyone who did not perform satisfactorily.

Another significant strike in shaping the attitudes of Mormon church leaders toward organized labor occurred at the Bullion-Beck mine at Eureka in the Tintic Mining District. The Bullion-Beck mine was owned by some of the leading authorities within the Mormon church. In January 1893, wages were reduced fifty cents a day, and the miners were required to trade at the company-owned store, saloon, and boarding house. The miners struck against these conditions and demanded that the company recognize their union. They succeeded in closing down the mine by January 17, but the company attempted to reopen the mine with imported strikebreakers. Riots broke out between the strikers and strikebreakers, and in early April, United States Marshal I. A. Burton subdued the riots by force. The Eureka Miners' Union was not recognized, but it persisted with enough strength to play a significant role in the organization of the Western Federation of Miners at Butte, Montana, in May 1893.

The investment by high-ranking church officials in the Bullion-Beck mine symbolized a change in the economic policy of church leaders. Before 1890, investment had been directed toward home industry, railroads, and cooperative enterprises. Because of the economic repercussions of the polygamy crusade of the 1880s, a definite economic policy did not emerge until after the Manifesto in 1890, which abolished polygamy. The new climate saw economic enterprises in Utah conform much more closely to the trends and philosophy of the entire country. This was enhanced by a new generation of church leaders whose background was primarily in business, in contrast to the first generation of leaders whose business experience was more limited and less intensive in nature. The strike and riots in Eureka during 1893, therefore, served as a personal example to those church officials who had invested in the Bullion-Beck mine of the potentially disruptive nature of organized labor.[14] Furthermore, radical unionism constituted a special affront to Bullion-Beck investors because it denied the very values by which they themselves had risen from the ranks of the poor.

The Eureka unrest was overshadowed by the national panic of 1893, which was severely felt in Utah. In Salt Lake City, the unemployment rate neared 50 percent in the spring of 1894, and demonstrations were staged to demand work.[15] Some 400 men joined the Utah Industrial Army with the intent of journeying to Washington, D.C., as part of a protest led by Jacob S. Coxey. However, only a handful of the Utah men actually reached their destination. The Utah union movement suffered as it commonly would in other periods of hard times. By 1896, however, recovery was apparent. Most unions either defunct or inactive because of the depression began to reorganize, and in mid–1896, the Utah Federation of Labor was founded.

Labor also made significant legislative gains during the mid–1890s. The Utah Constitution, which became effective when Utah was granted statehood on January 4, 1896, contained several important items regarding labor. Article XVI entitled "Labor" contained seven sections; four sections provided for a Board of Labor, Conciliation, and Arbitration, an eight-hour day on public works, and health and safety legislation in factories, smelters, and mines. Three sections prohibited employing women or children under fourteen years of age in underground mines, the contracting of convict labor, the political and commercial control of employees by employers, and the exchange of blacklists.[16] Two sections of Article XII entitled "Corporations" were also concerned with organized labor. Section Sixteen prohibited corporations and associations from importing armed men to preserve the peace or suppress domestic troubles without lawful authority, and Section Nineteen tried to discourage blacklisting practices.

The 1896 state legislature passed an act incorporating most of the ideas contained in Article XVI of the state constitution. It also extended the provisions of the eight-hour day to cover workers in mines and smelters, provided for the appointment of a coal mine inspector, and defined in detail safety regulations to be followed in mining operations. The eight-hour law was challenged in *Holden* v. *Hardy*; however, the law was upheld in a landmark decision by the United States Supreme Court.[17]

With the new state government and constitution launched in an era of relative prosperity, the period between 1901 and 1910 was seen as "the active years in the Utah labor movement."[18] The state's population increased from 276,749 in 1900 to 373,351 in 1910. The population of Salt Lake City rose from 53,351 to 92,777, an increase of 73.3 percent.[19] By the end of the decade, forty-seven Salt Lake City union locals and twelve Ogden locals were organized. In addition, the railroad unions appeared well established, and the Western Federation of Miners had organized in the major metal-mining communities of the state.

The Utah union movement was greatly strengthened by stronger ties between local unions and the national organizations through direct affiliation and through the efforts of national organizers and visitors to the state such as labor spokesmen Samuel Gompers, president of the American Federation of Labor, and Eugene V. Debs, the popular Socialist candidate for president of the United States.[20] In 1904, the Utah Federation of Labor held the first statewide convention for all labor unions in Utah.

The drive by organized labor after 1900 met strong resistance by Utah employers. The labor organizations threatened the economic supremacy of the employers, who met the challenge by establishing a Citizens Alliance. Through this alliance and other means, employers protected their interests from the unions and mounted an aggressive counteroffensive that kept organized labor on the defensive until the 1930s.

The period between 1901 and 1904 proved especially restless as strikes were called by the carpenters, cooks and waiters, stonecutters, teamsters, ironworkers, bakers, machinists, telephone linemen, Western Union messengers, electrical workers, and coal miners.[21]

During the first decade of the twentieth century, most strikes were called by the craft unions, but this changed during the second decade as the theater of action shifted from the conservative labor elements to those more radical in nature. In 1910, two bombs exploded at the construction site of the Hotel Utah. The bombings were eventually connected to J. E. Munsey, business agent for Local No. 27 of the International Association of the Bridge and Structural-Iron Workers. The Hotel Utah bombings were the first of several terroristic endeavors that would gain national notoriety for the Bridge and Structural-Iron Workers Union. Later that year, on October 1, 1910, a bomb explosion at the Los Angeles Times building killed twenty people and injured seventeen. J. J. McNamara, secretary-treasurer of the union, and his brother J. B. McNamara, a member of the union's executive board, were arrested and confessed. Mormon leaders regarded the bombings of the church-controlled Hotel Utah together with problems in other parts of the country as evidence of the danger of radical unions.

The election of 1912 in Utah seemed to indicate that organized labor was on a collision course with what many considered the sacred principles of capitalism. Despite the efforts of the more conservative faction of the Utah Federation of Labor, the statewide organization at its seventh convention, held in September 1911, voted to support the Socialist party during the upcoming election.

> Whereas in the entire history of the American Labor Movement the police, the military, and the judicial department of our state and national government have been used in the interest of capital rather than of labor, in the interest of property rather than of man. . . . Whereas, this hostility too has been shown by governments holding their power by the will of each and all of the old capitalistic supported parties and whereas, there is but one political party in the field today which stands forever with labor as against capital and for the final complete emancipation of the entire working class, and whereas, that party is known as the Socialist Party, be it therefore resolved, that we . . . hereby endorse the said Socialist Party . . . that we call upon all members of organized labor in the State of Utah to study the principles and aims of socialism and its representative party . . . to lend their aid to this political party . . . that as a state organization we aid in the propaganda of socialism that we may hasten the day when the emancipation of the working class from the bonds of wage slavery shall be proclaimed in America and throughout the world.[22]

The support of the Utah Federation of Labor proved ineffectual, and the Socialist party received only 9,000 or 8 percent of the 112,300 votes cast.

During the troubled years between 1911 and 1913, strikes centered in the railroad, smelting, and mining industries. On October 1, 1911, railroad workers in Salt Lake City and Ogden joined a national strike against the Harriman railway system. The strike lasted until early in 1914 and was characterized by conflicts between strikers and strikebreakers. Benefit balls were held by local unions in Salt Lake City and Ogden to raise funds for the striking railroad workers, and members of the Western Federation of Miners (WFM) at Bingham donated $1,000 to the strike cause. Nevertheless, the strikers were defeated despite their thirty-month strike.[23]

During 1912, the Western Federation of Miners called two strikes to secure higher wages and recognition of the union. A strike by the men at the American Smelting and Refining Company smelter in Murray began on May 1, 1912. The strike lasted six weeks and was finally broken with the importation of Cretan strikebreakers. Relations had hardly been restored, however, when the Bingham strike began on September 17, 1912, as 4,000 workers left the mines. This time the strikers' ranks included the Greek miners whose main grievance was their forced subservience to the labor agent, Leondias Skliris. Strikebreakers were imported from the East; and in a battle between the strikers and the strikebreakers and deputies hired by the Utah Copper Company, two men were killed. The strike lasted nearly a year, and in the end, the copper company proved victorious as many WFM strikers went back to work without union recognition by the company.

Organizers of the Industrial Workers of the World (IWW or Wobblies) were present in Utah shortly after its establishment in 1905. However, it was not until the summer of 1913 during a strike against the Utah Construction Company camps located along the railroad between Helper and Tucker, that the Wobblies in Utah tested their strength. The unsuccessful strike was followed by a free-speech campaign, confined primarily to Salt Lake City. At one of the meetings, a riot broke out when James F. Morgan, an IWW spokesman, was attacked by Axel Steele, a Utah Copper Company employee in charge of deputy sheriffs at Bingham.[24]

The Tucker strike and the IWW free-speech movement set the stage for Utah labor's most famous episode—the trial and execution of IWW songwriter Joe Hill. A native of Sweden, Joe Hill immigrated to the United States in 1902 and joined the IWW in 1910. Hill's primary contributions to the IWW movement were the songs he wrote, including "The Preacher and The Slave," "There is Power in the Union," and "Workers of the World, Awaken!"[25]

In 1913, Joe Hill left California for the East but stopped to work in the mines of Utah. While in Utah, he was arrested for the murder of a Salt Lake City grocer, John G. Morrison. Circumstantial evidence, especially a gun wound in Hill's arm that he refused to explain satisfactorily, led to his conviction; and despite an international protest, Hill was executed by a firing squad on November 19, 1915. The IWW effectively capitalized on Hill's death, charging that he had been sacrificed by the Utah Copper Company, the Utah Construction Company, and the Mormon church as part of an all-out attack against organized labor. In death, Joe Hill became American labor's best-known martyr. His last wish, to have his body hauled to the Wyoming border because "I don't want to be found dead in Utah," branded Utah, in the eyes of much of the working class, as a stronghold of reactionary anti-unionism. The IWW staged an impressive funeral for Joe Hill in Chicago on November 26, 1915, after which his body was cremated and the ashes placed in envelopes and distributed to IWW locals in every state but Utah. Envelopes were also sent to South America, Europe, Asia, South Africa, New Zealand, and Australia.[26]

The prosperity of World War I provided labor with an opportunity to expand its Utah organization; however, the most significant gains were made in the area of labor legislation. In 1916, Simon Bamberger, identified as a "friend of labor," gained the governorship of the state, and several labor leaders were elected to the state legislature.[27]

The 1917 legislature passed several important labor measures including: the establishment of a State Industrial Commission, a workmen's compensation law, and an act entitled "Bettering Conditions of Labor." The act guaranteed workers the right to organize into unions for the purpose of reducing work hours, increasing wages, and bettering working conditions. In order to achieve their objectives, labor unions had the right to strike and carry out peaceful picketing.[28]

The postwar years of 1919 and 1920 were among the most restless in American labor history. The triumph of the Bolsheviks in Russia terrified many Americans, who became convinced that the United States might be the next victim of the "atheistic radicals." The Seattle general strike, nationwide strikes in the coal and steel industries, the Boston policemen's strike, and a rash of bombings reinforced the widespread fear that America would soon be engulfed by the Bolshevik menace.

There were several manifestations of leftist activity in Utah. A Workers, Soldiers, and Sailors Council of Salt Lake City and County was organized in 1919 and dedicated its efforts to overthrowing capitalism and restoring political and economic power to the workers. In Park City miners struck on May 6, 1919, demanding a six-hour day, $5.50 in wages per day, and no discrimination because of union membership. The strike was the most pervasive in the fifty-year history of the Park City Mining District. Also in May 1919, the cooks and waiters of Salt Lake City went on strike. Rumor spread that the 4,500 members of the Salt Lake Building Trades Council would leave their jobs in sympathy with the cooks and waiters, thus bringing a general strike to Salt Lake City. In the fall of 1919, federal troops from Fort Douglas, Utah, and Camp Irwin, California, were sent to Utah's coal fields in anticipation of violent support by Utah coal miners of the nationwide coal strike called for November 1. At the annual convention of the Utah Federation of Labor, held September 1919 in Salt Lake City, the radical element in Utah labor succeeded in controlling the convention, despite the resistance of the Salt Lake Typographical Union and other craft unions in the state. Under the leadership of these radicals, the convention endorsed the Soviet government of Russia and called for the abolishment of the contract system between workers and employers.[29]

Vigorously responding to the challenge by organized labor, Utah employers regained the offensive through the "American plan," or open-shop system. The plan insured that "employment shall not be refused to any workman on account of his affiliation or non-affiliation with any labor organization."[30] The right of the employer to hire and fire any employee under any circumstances was also guaranteed.

The Mormon church became the most influential group supporting the open-shop movement in Utah. In 1919, church President Heber J. Grant outlined his position toward organized labor and expounded on the conflict between the closed shop and individual free agency.

> I want to say that I am perfectly willing that men shall join labor unions, that they shall band together for the purpose of protecting their rights, provided they do not interfere with the rights of other people. Life, liberty and the pursuit of happiness belong to all people in the United States, according to the laws of our country . . . and I say that, to my mind, a provision in a labor union is all wrong that favors boycotting and the laying down of tools or the quitting of employment because a non-union man obtains employment while exercising his God-given right to stay out of a union. Men who have that kind of a rule have a rule that is in direct opposition to the laws of God. . . . How much love is there in starving your neighbor because he will not surrender his manhood and

> his individuality, and allow a labor union to direct his labor? Mighty little love, mighty little of the gospel of the Lord Jesus Christ in any such rule! I hope to see the day when no Latter-day Saint will join a union unless the union eliminates that clause from its rules. I am not going to ask them to leave their union, I am not going to lay it down that they must, that it is the mind and will of the Lord for them to leave a union . . . but I cannot see how a Latter-day Saint who is a member of such a union can get down on his knees and pray for God to inspire and bless him. . . .[31]

Most Utah union members sympathized with Grant's idealistic stand on free agency but recognized that in the down-to-earth struggle for better conditions such a position would be used by employers to weaken and break the union. In Utah, the building industry was the first target in the open-shop campaign. The building trades unions were unable to withstand the onslaught, and in August 1920 a committee representing each building craft reported to the Salt Lake Building Trades Council that the various crafts had little choice but to accept the open-shop idea.[32] Reflecting the anti-union sentiment implied in the open-shop doctrine, the Utah State Legislature passed a "right-to-work" act during the 1923 session.

In addition to the open-shop campaign, several other factors retarded the growth of organized labor during the 1920s. A depression in 1920 and 1921 left many workers unemployed. The oversupply of labor increased when employers, in an effort to force acceptance of the open-shop idea, encouraged workers from outside the state to come to Utah.[33] In addition, organized labor witnessed several changes in leadership as many experienced labor leaders died during the 1920s.[34] Local unions proved powerless to prevent several wage reductions instituted during the decade. The *Utah Labor News*, labor's mouthpiece in Utah, became defunct in 1924, and most union activity was concerned only with the internal business of the local unions.[35]

Labor in Utah followed the national trend throughout the 1920s and into the Great Depression. By mid–1932, unemployment in Utah reached approximately 28 percent.[36] The election of Franklin D. Roosevelt and the subsequent passage of the National Industrial Recovery Act (NIRA), which guaranteed employees (in Section 7a) the right to collective bargaining through representatives of their own choosing, proved to be not only the most important milestone in Utah's labor history, but in the entire nation as well. Under the protection of the New Deal administration, union membership in Utah rose from 1,603 in November 1934 to 4,389 two years later.[37] In addition, the new members came from industries, such as coal and metal mining, which had not been previously organized.

The year 1935 represented an important change in Utah's labor history. The New Deal's labor policy gave organized labor a legitimacy that it previously lacked. With the unionization of Utah's industrial workers, the influence of the craft unions in the development of Utah labor policy and philosophy declined. Utah's coal miners played a major role in this change because in no other industry or trade was the opposition to unionism so strong as in Utah's coal fields. By 1935, coal miners represented approximately 40 percent of the membership affiliated with the Utah State Federation of Labor.[38]

The Coal Industry

The beginnings of Utah's coal industry extend back to the arrival of the Mormon pioneers in the Salt Lake Valley and to Brigham Young's push for the development of a self-sufficient coal and iron industry. Such industries, the leader maintained, were essential to successful Mormon colonization of the West. In comparing England and Spain, Young emphasized the importance of coal and pointed to "the wealth, strength, and glory of England, growing out of her coal mines, iron, and industry; and the weakness, corruption and degradation of . . . Spain growing out of her gold, silver . . . and idle habits."[39] Young did not live to see his dream of a successful coal industry realized, although he saw it begin and falter in three regions—Iron County, the Sanpitch Mountains, and the Coalville area east of Salt Lake City in Summit County. Nevertheless, at the time of Young's death in 1877, development of mines in eastern Utah began, marking the end of an illusive three-decade pursuit of a viable coal-mining district within the Mormon kingdom.

The Iron Mission

The search for coal began in 1847 when exploring parties visited the valleys surrounding the Great Salt Lake, looking for suitable agricultural land and, among other items, coal. These exploring parties did not discover coal immediately. In 1849, fifty men, led by Parley P. Pratt were sent to explore southern Utah and located a rich deposit of iron surrounded by thousands of acres of cedar trees, which, the explorers reported, could be used to smelt the iron ore. Brigham Young responded to the opportunity to develop this important resource; and the following year an advance party of 119 men, 30 women, and 18 children journeyed 250 miles south from Salt Lake City and established the settlement of Parowan as the vanguard of what came to be the Iron Mission.

Coal was discovered in the nearby mountains, and under the careful and enthusiastic nurturing by Mormon leaders, "by the fall of 1851 virtually every man in Utah territory who understood the working of coal or iron had been sent to Parowan."[40] Although the iron ore was of high quality, the nearby coal was unfortunately weak and friable with a high-sulfur content; coke could not be produced from it. Good-quality iron was produced by using charcoal and wood as fuel, but the expense of providing large quantities of wood required for the smelting process proved that it was not an acceptable substitute for coked coal.[41] The pioneer iron industry failed primarily because of the smelting problems. Despite several subsequent attempts, it was not until 1924, when the Columbia Steel Corporation erected a plant at Ironton in Utah County, that ore from Iron Mountain was successfully used to produce large quantities of iron. The iron ore was transported 200 miles by the Union Pacific Railroad to Ironton where it was smelted using a high-quality coke from Carbon County.[42] For seventy-five years, the lack of a suitable coking coal prevented the development of the iron deposits discovered by the Southern Utah Expedition in 1849.

Wales, Sanpete County

As the experiments in Iron County with coal for coking purposes proved unsuccessful, a new coal source was discovered in 1854, 150 miles to the north in the Sanpitch Mountains on the western edge of Sanpete Valley. When Brigham Young visited the valley, he was given pieces of coal by a local Indian named Tabiona. Anxious to determine the quality and quantity of the Sanpete coal, Young sought anyone familiar with coal mining. John E. Rees, a Mormon convert who began work in the coal pits of his native Wales at the age of eight, responded to the inquiry and reported that John Price, a settler in Manti, was also knowledgeable about coal mining and possessed the necessary tools.

At Brigham Young's request, Tabiona guided the two men to the deposit located about twelve miles northwest of Ephraim. Price and Rees commenced mining coal and within a short time were joined by several families; the settlement of Coalbed was established by the end of 1854.

Brigham Young visited the coal discoveries in Sanpete Valley during the spring of 1855 and became convinced that the coal beds would provide the territory's residents with a cheaper and better source of fuel. Early that summer, Coalbed shipped several wagonloads of coal to Salt Lake City.[43] In 1859, the coal-mining effort received a substantial boost by the arrival of several families of Welsh converts. However, the mines were not fully developed because of the agricultural necessities that limited coal mining to the winter months, and with continued Indian difficulties, the settlement was abandoned in 1866. Resettled in 1869, Coalbed was renamed Wales in honor of the homeland of most of the town's residents.

The local owners of the coal deposits did not have sufficient economic resources to increase the production and distribution of coal from the mines much beyond a few hundred tons annually, and in 1872 the mines were sold to a company organized by John T. Lynch and C. C. Perkins of Salt Lake City. They hired 200 coal miners, built twelve coke ovens, and installed crushing and washing machinery in an effort to expand the production and market for the Sanpete coal. However, small seams of rock interlaced the four-foot vein of coal, and despite the large quantities of coke produced, it proved unsatisfactory for smelting purposes.[44] Within two years, the Lynch-Perkins Company failed, and Simon Bamberger, later governor of Utah, took up the cause of Sanpete coal. Along with Charles Bennett, a prominent Salt Lake attorney, Bamberger traveled to England to solicit capital to develop the mines. They successfully organized the Central Pacific Coal and Coke Company. Sir Henry Tyler, president of the Grand Trunk Railroad of Canada, headed the new company. In 1880, the company constructed a narrow-gauge railroad, twenty-seven miles long from the Utah Southern Railroad at Nephi to Wales, and new mines were opened in Pete's Canyon south of Coal Canyon.[45]

The development of mines at Wales after the initial attempt by Mormon converts was made primarily by non-Mormon interests. Although church leaders encouraged further attempts to develop the Sanpete coal fields, by 1870 they looked to the mines of Summit County as a source of coal for the Salt Lake Valley. Meanwhile, the Sanpete mines met with uneven success, prospering during World War I only to fade to a local wagon mine shortly after the end of the war.

The Coalville Coal Fields

Utah's territorial leaders were anxious to encourage the discovery of a workable vein of coal within a reasonable distance of the major population center of Salt Lake City. In January 1854, the Utah Territorial

Legislature offered a reward of $1,000 to any resident of the territory who would "open a good coal mine, [the coal vein] not less than eighteen inches thick, within forty miles of Great Salt Lake City, in any accessible position, and that can be profitably worked. . . ."[46]

Although coal was discovered in the Sanpitch Mountains during 1854, it was approximately 120 miles from Salt Lake City. Four years later in 1858, Thomas Rhoades discovered an outcropping of coal while hunting in the vicinity of Chalk Creek, about forty-five miles east of Salt Lake City. An 1835 convert to Mormonism and veteran of the California gold rush, Thomas Rhoades was reportedly the custodian of a fabulously rich gold mine, which, according to tradition, had been first mined by the Spaniards. The legend maintained that after the Spanish were driven away by the Indians, the mine was preserved until Chief Walker offered to give Brigham Young all the gold he needed as long as the source and location remained secret. Only one man, Thomas Rhoades, was permitted to secure the gold. After locating the outcropping of coal near Chalk Creek, Rhoades removed a piece with a butcher knife and presented it to Brigham Young. The next year F. B. Meek and E. Englemann, two members of Captain James S. Simpson's 1859 expedition, also reported seeing coal in the Chalk Creek-Weber River area.[47]

An agricultural settlement had been established at the mouth of Chalk Creek in 1858, about the time that Thomas Rhoades discovered the nearby outcroppings; however, it was not until 1860 that Brigham Young sent John Muir and Samuel Fletcher to investigate the discovery. Unable to find the coal reported by Rhoades, the two men accidentally discovered a ten-foot-thick coal vein on Grass Creek in the canyon north of Chalk Creek.[48]

John Spriggs, who arrived at the Chalk Creek settlement in June 1859, located the Thomas Rhoades find and attempted to mine coal there. In 1860, he and William Kimball claimed the $1,000 reward offered in 1854, but a committee of the territorial legislature refused to honor the claim on two points: the mine was more than forty miles from Salt Lake City, and the quality of the coal was inferior.[49]

The find by Muir and Fletcher proved more suitable; and the Old Church mine was opened on Lion's Hill by Daniel H. Wells, Bryant Stringham, and Stephen Taylor. Coal was hauled by wagon to Salt Lake City where it sold for between thirty-five and forty dollars a ton.[50] Other mines were opened in Allen's Hollow, Spring Hollow, and Dexter's Hollow adjacent to Chalk Creek. The most important of these mines was the Wasatch mine.

With the completion in 1869 of the Union Pacific line five miles north of Coalville at Echo Junction, the railroad effectively controlled the Salt Lake City market. In a classic example of ruthless monopoly and cut-throat business practices, the Union Pacific Company initiated exorbitant short-haul rates and other discriminatory practices against the Utah coal operations in an effort to control the market for its Wyoming coal mines. Utahns made repeated attempts to break the monopoly of the Union Pacific Railroad and stimulate further development of the Coalville coal fields. The completion of the Denver and Rio Grande Western Railroad from Denver to Salt Lake City in 1883 provided the breakthrough Utahns had long hoped to see, and Mormon and territorial leaders looked to the new railroad for salvation from the high coal prices. As Utahns turned to the newly opened eastern Utah coal fields, the Coalville operations assumed a secondary role in Utah's coal production.

Connellsville

The John W. Gunnison expedition in 1853 made the first recorded discovery of coal in eastern Utah. Surveying a possible transcontinental railroad route along the 38th and 39th parallels, the expedition joined the Old Spanish Trail at the Green River crossing and made its way through Castle Valley. On October 11 while camped near the eastern slope of the Wasatch Plateau, approximately three miles east of the present community of Emery, the expedition discovered coal. Lieutenant Edward G. Beckwith reported: "Specimens of coal were brought in from the hills near the camp. Captain Gunnison and Dr. [James] Shiel differing in opinion as to quality."[51]

Because of the isolated location of the Gunnison discovery and the hopes that Wales and Coalville would produce enough coal to meet territorial needs, the brief report was forgotten. After the coking efforts at Wales failed, promoters looked to the mountains east of the Sanpete settlements for sources of good coking coal. In May 1874, the Fairview Coal Mining and Coke Company was organized.[52] The company opened a mine in Coal Canyon near the head of Huntington Canyon and built eleven coke ovens. A small mining camp with big aspirations, Connellsville, named for the large coking center in Pennsylvania, was established, and coking efforts were carried out for approximately three years. Wagons hauled coke to Springville, but transportation costs and the questionable quality of the coke doomed Utah's Connellsville to a short existence.

Pleasant Valley

Eastward nine miles, across the mountain from Connellsville, is Pleasant Valley, which today holds the community of Scofield and the fisherman's mecca of Scofield Reservoir. The area served exclusively as pasture land until 1875 when the Pleasant Valley Coal Company began development of the coal resources. The Pleasant Valley Coal Company was headed by Milan O. Packard of Springville and a board of directors that included Martin Crandall also of Springville; Warren D. Child of Ogden; and Frank Pritchett, Lark Thompson, and John Thompson of Fairview.[53]

The reports of coal in the Pleasant Valley area created a minor rush to claim the coal lands, and during the winter of 1875–76, John Nelson and Abram Taylor spent several months in the isolated mountain valley securing the Pleasant Valley Coal Company claims. The town, developed after the mines were opened, was named Winter Quarters in commemoration of the two men's winter vigil.[54]

During the spring of 1876, the Pleasant Valley Coal Company began constructing a wagon road from Springville up Spanish Fork Canyon to its Pleasant Valley coal lands.[55] The following year Peter Moran and fourteen other men crossed the mountains from Huntington Canyon to Pleasant Valley and opened what became known as the Winter Quarters No. 1 mine. The men had not planned to stay the winter but were snowed in. Forced to remain until February 1878, their supplies exhausted, the Moran group struggled about twenty miles through deep snows to a lonely sawmill camp named Tucker at the foot of Soldier Summit on present-day U.S. Highway 6.[56]

Directors of the Pleasant Valley Coal Company acted quickly to improve the inadequate transportation system. During the first months coal was hauled by wagon to Springville and Provo, a trip requiring four days in good weather, but impassable in winter. Organized in 1877, the Pleasant Valley Railroad Company began constructing a narrow-gauge line from Springville to Winter Quarters, which required most of 1878 and 1879. In October 1879, the railroad, known as the "Calico Road" because Milan Packard paid workers part of their wages in merchandise from his Springville store, was completed to Winter Quarters. However, as the Denver and Rio Grande Western Railroad (D&RGW) approached Salt Lake City from Denver in March 1881, rumors began to circulate that the Pleasant Valley Railroad would be acquired by the larger company. In June 1882, the D&RGW purchased the line.[57] Because the grade of the Pleasant Valley Railroad was excessively steep and the narrow-gauge track inadequate, in 1884, the D&RGW built a branch line from the main line at Colton, up Fish Creek to Pleasant Valley.

One of the important questions about the Pleasant Valley coal was the nature of its coking qualities. Shortly after the mines were opened, twelve coke ovens were erected, and extensive coking tests were undertaken. However, like the earlier discoveries in Utah, the coal proved unsatisfactory for coking purposes.

In 1880, David Williams, the Mormon bishop in Winter Quarters, leased the Winter Quarters mine from the Pleasant Valley Coal Company; work at the company's newly opened Mud Creek mine was halted. Later, when the Pleasant Valley Coal Company was acquired by the D&RGW in 1883, work resumed at the Mud Creek mine despite the protests of Bishop Williams, who charged that the development infringed his lease and that the market could not handle the production from both mines.[58]

After the Mud Creek mine was reopened in 1883, Chinese miners were imported as laborers. They remained until 1885 when hostility against the Chinese climaxed during the riots in Rock Springs, Wyoming, and the Chinese miners were forced out of coal mines in the West.[59]

In 1885, David Williams's lease of the Winter Quarters mine expired, and the Pleasant Valley Coal Company assumed operation. Under the direction of the company's production superintendent W. G. Sharp, production averaged 7,000 to 8,000 tons a month.

The approach of the D&RGW and the opening of the Pleasant Valley coal fields jeopardized the Union Pacific Railroad's monopoly of the Salt Lake City coal trade. In response, the Union Pacific encouraged the Utah Central Coal Company to penetrate the Pleasant Valley area in 1882. The company found a twenty-eight-foot vein of coal in the mountain just east of Scofield. The mine was opened by John Fife and operated until 1884 when the tipple, a structure used to process coal, caught fire on New Year's morning and set the coal mine on fire. A miner named John McLean and his son were killed, and the mine was permanently sealed after attempts to extinguish the fire proved unsuccessful. A year later a second mine was opened.[60] In 1887, the Utah Central Coal Company induced W. G. Sharp to leave the Pleasant Valley Coal Company and take a job as superintendent of its mines in Pleasant Valley, a position he held until 1892. In November 1890, the Union Pacific Coal Company took control of the former Utah Central coal mines and ran them for a time. It was anticipated that the mines would provide coal for the Utah Central Railroad, which ran from Ogden to the mining town of Frisco in Beaver County.[61] The Utah Central Railroad planned to build a road into Scofield and haul its own coal, but the road was never built. The coal from the Union Pacific mines was hauled by the D&RGW. The arrangement was less than ideal for the Union Pacific because of the excessive freight rate charged by the D&RGW. The rate amounted to $1.25 a ton for the sixty-seven mile trip from Scofield to Provo with the Union Pacific Railroad required to provide its own coal cars. The Union Pacific Railroad ceased operating the mine in 1897.

The Utah Fuel Company

A second coal company, known as the Utah Fuel Company and incorporated in New Jersey, was organized by the D&RGW in 1887 to operate the Sunnyside and Clear Creek mines in Utah and the Sommerset mine in Colorado. The Pleasant Valley Coal Company, owned by the D&RGW, continued to operate the Winter Quarters and Castle Gate mines until the company consolidated with the younger Utah Fuel Company in 1899.[62]

Because the coal from the Pleasant Valley area had proven unsuitable for the production of coke, coal developers continued their search for the illusive vein of good coking coal. Developments after 1888 in Castle Gate on the Price River near the mouth of Price Canyon were more successful. Coal mined in Castle Gate proved suitable for coking purposes; in 1889, several coke ovens were erected. By 1896, 104 coke ovens had been constructed, and coke production amounted to 20,448 tons for the year.[63] The Coal Mine Inspector's Report for 1897 noted, "This is by far the most extensively and systematically operated coal mine in the state."[64] Within five years, an additional 100 coke ovens were in operation, and coke production from the Castle Gate ovens amounted to approximately 5,000 tons a month.[65]

But even Castle Gate lacked totally satisfactory coal for coking purposes, and by 1901 most coke produced there was from coal mined at the Utah Fuel mines in Sunnyside, some forty miles to the southeast. Whitmore Canyon, where the Sunnyside mines were opened in 1899, was initially settled by George C. Whitmore during the 1870s. A powerful man in southeastern Utah, Whitmore operated a cattle ranch on Grassy Trail Creek, the main water source for what has since become known as the East Carbon area. The expansion of the D&RGW into the Sunnyside area through its subsidiary, the Utah Fuel Company, was controversial. Utah Fuel Company officials attempted to buy the Whitmore ranch with its all-important water rights. Whitmore refused the company's offer, and it, in turn, rejected his counteroffer of $90,000.[66] The coal company was forced to dig wells and develop water sources inside the mine.

More questionable was the acquisition of coal lands by the Utah Fuel Company. In 1904, suits were filed against the Utah Fuel Company for illegal acquisition of public coal lands.[67] A grand jury investigation led to the indictment of H. G. Williams, general manager of the Utah Fuel Company, and of several other company employees for conspiracy to defraud the federal government. According to testimony before the grand jury, the agents of the Utah Fuel Company, primarily Robert Forrester, the company geologist, had hired young men known as "dummies" at fifty dollars plus expenses and a three-dollar per diem. They were to file on and purchase, with funds provided by Utah Fuel, coal lands on which entries had been made under Forrester's direction. The lands were then transferred to the coal company.[68] Also in question was the acquisition of some coal lands as "agricultural lands" for which the entry fee was only $1.50 an acre instead of the $20 an acre for coal lands.[69] Much of the testimony of witnesses for the government, including that of George Whitmore, substantiated the contention that the lands were known only for their coal and not for their agricultural value.[70]

The litigation lasted for five years and involved other acquisitions outside the Sunnyside area. Finally, the Utah Fuel Company was forced to settle. The company was fined $8,000; and was required to pay $192,000 for coal extracted from land fraudulently acquired in Salina Canyon in Sevier County; to return the Salina Canyon coal properties; and to forfeit the $14,400 paid in entry fees.[71] The company retained possession of their Carbon County lands; however, in a subsequent decision made in July 1909, it was required to pay an additional $73,000 for coal extracted.[72]

Although the Utah Fuel Company lost a total of $287,400 plus the attorney's fees and the land in Salina Canyon valued at $100,000, it continued to increase production at its Carbon County mines. The company found it difficult to obtain sufficient American miners and increasingly looked to new arrivals from Italy, Greece, and the Balkans to fill the labor shortage in the coal mines.

The primary effort during the first seven years of the twentieth century centered on the development of the coking industry. In 1900, coke production stood at 50,620 tons; however, with the construction of 200 coke ovens at Sunnyside, coke production reached 139,792 tons in 1902. By 1907, an additional 450 coke ovens had been constructed at Sunnyside, and the total output of coke from Carbon County was 324,692 tons for the year. The panic of 1907 had a significant impact on Utah's coal industry. In Carbon County, coal production declined from 1,816,133 tons in 1907 to 1,606,853 tons in 1908. Coke production dropped to only 134,195 tons, a reduction of 190,500 tons from the previous year.

The Independent Coal Companies

The coal land fraud cases against the Utah Fuel Company and a continuing coal shortage in Salt Lake City encouraged the organization of independent coal companies that sought to develop further the resources of the eastern Utah coal fields. The first of these companies was called appropriately the Independent Coal and Coke Company. This company opened mines and founded the town of Kenilworth, a few miles east of Helper. A railroad line connected with the D&RGW at Helper on which Shey engines powered the coal trains because of the steepness of the grade. The Independent Coal and Coke Company, in direct competition with the railroad's Utah Fuel Company mines, was placed at the mercy of the D&RGW. Consumers feared the D&RGW would meet the competition in the same manner as the Union Pacific Railroad had handled coal from the Coalville area during the 1870s. An editorial in the *Intermountain Republican*, warned the D&RGW that if it attempted to discriminate against the independent coal companies, the public would establish a railroad commission, which would work to secure legislation regulating the railroads.[73]

The D&RGW reacted much more judiciously in dealing with the independent coal companies than had the Union Pacific. Instead of trying to destroy competition from the other coal mines, the Utah Fuel Company sought to reach an agreement with the independent producers for a standard price on coal. The cooperation among coal producers soon led to accusations of exorbitant price-fixing. Three years after the *Intermountain Republican*'s warning against discrimination, the Salt Lake City newspapers headlined charges against a coal trust and demands for an investigation by the Interstate Commerce Commission.[74] In November 1910, the *Salt Lake Telegram* concluded, "When the ghosts of Jay Gould and Sidney Dillon look over the daily market reports of Salt Lake City, they must in muffled tones whisper to each other: 'And there were men in our day in Salt Lake who thought we were grafters.' "[75]

Meanwhile, the Hiawatha area southwest of Price was opened by two additional independent coal companies, the Consolidated Fuel Company and the Castle Valley Fuel Company with mines at Hiawatha, Black Hawk, and Mohrland. By 1910, the two companies succeeded in running the Castle Valley Railroad from Price to their mines at Hiawatha and Mohrland. In 1912, the mines were purchased by the United States Fuel Company, a subsidiary of the United States Smelting, Refining, and Mining Company which had interests in the copper smelting industry then flourishing in Salt Lake City. The priority for the new owners was the construction of a railroad to the mines since the original railroad, built in 1910, proved "poorly . . . constructed, with impracticable grades, lacking in equipment, and incapable of . . . handling the output of the mines, to say nothing of making possible increased output and new development."[76]

Instead of relying on the D&RGW to handle the coal cars after they reached the main line at Price (as was the arrangement with the Castle Valley Railroad), the new Utah Railway Company planned to build a separate line paralleling the D&RGW tracks to Spanish Fork where it would then connect with the San Pedro, Los Angeles, and Salt Lake Railroad, part of the Union Pacific system. However, once again the D&RGW was willing to accommodate the new line. When it proposed to reconstruct and relocate its lines as part of a double tracking of the road between Thistle and Castle Gate, it offered the Utah Railway trackage rights on the double track and suggested the new road be built by the railway between Thistle and Provo and be used jointly with the existing D&RGW track. The Utah Railway wisely agreed to the proposal, and construction of the new road between Hiawatha and Castle Gate commenced in October 1912 and reached completion two years later.[77]

While the United States Fuel Company was acquiring possession of the coal fields in the Hiawatha area, Jesse Knight, a Mormon businessman from Provo, began developing coal lands in the Spring Canyon area. In the summer of 1912, he acquired 1,600 acres of coal lands in the canyon. In September 1912, A. E. Gibson was hired as superintendent, and construction began on the company town named Storrs, in honor of

George A. Storrs, an associate of Jesse Knight. Actual coal production began in February 1913, when a railroad line was sufficiently completed to handle coal shipments.[78]

The second mine in Spring Canyon followed closely on the heels of the Jesse Knight mine. The Standardville mine was developed by the Standard Coal Company in 1913. The company was organized by F. A. Sweet after he sold out his interests in the Hiawatha area to the United States Fuel Company the previous year.

In 1914, the Liberty Fuel Company opened the Latuda mine in Spring Canyon, and in 1916 two additional mines at Peerless and Rains were developed. The opening of the Spring Canyon mines occurred just as World War I created a tremendous demand for coal. During 1916, production at the Standardville and Storrs mines equalled that of the older mines at Winter Quarters, Clear Creek, Kenilworth, Hiawatha, and Black Hawk.[79]

In 1916, the Wattis mine, located a few miles north of Hiawatha, was opened by Ogden businessman W. H. Wattis, for whom the camp and mine were named. In 1919, Lion Coal Company acquired the mine. Because of the excellent coking quality of the coal in the Sunnyside area, the Columbia mine was opened three miles southeast of Sunnyside in 1922. The mine became an important part in the development of Utah's first successful steel mill constructed by the Columbia Steel Company at Ironton in Utah County.

The last important coal-mining area opened in Carbon County was in 1923 in the Gordon Creek area, south of Spring Canyon. Here, mines were developed by the Sweet Coal Company, National Coal Company, and Consumers Mutual Coal Company.

The rapid expansion and development of Utah's coal industry represented a significant investment by entrepreneurs inside and outside the state. Nevertheless, the developments proceeded at a significant cost in human lives. The threat of death or serious injury was a constant companion to anyone who entered the mines, whether an experienced miner from the coal pits of Wales, a Mormon farmer and part-time coal miner from Emery or Sanpete County, or a young southeastern European immigrant drawn to the mines with the promise of earning enough to return home to enjoy previously unknown financial security. The coal mine inspectors' reports and newspaper articles from the time testify that accidents and death were inseparable from coal mining—though the total of hundreds of random individual accidents was a quickly forgotten summary of danger in the coal mines. However, the deaths of at least 200 men within a few minutes in one mine could not be ignored. For Utah, the Winter Quarters disaster, also known as the Scofield disaster, was such an event, and served as the symbolic tombstone for all who died extracting coal. The rallying cry "Remember the Winter Quarters Disaster" chilled the blood and set the resolve of miners seeking redress of grievances and equitable pay for dangerous work.

Chapter II

The Winter Quarters Mine Disaster

On May 1, 1900, at least 200 men lost their lives in what was then the most disastrous mine explosion in the history of the United States.[1] The explosion at the Winter Quarters mine—located in a narrow, remote canyon more than eight thousand feet above sea level, and a mile west of Scofield in Utah's Carbon County—affected the course of labor not only in Utah but also in other parts of the country. The disaster clearly reminded workers of the ultimate sacrifice they might be required to make for their employer. The aftermath brought intense human suffering to the families and friends of the dead. The emotional shock for those who lost husbands, sons, and brothers was as great as the financial upheaval created by the loss of the family breadwinner. In a time when industrial insurance was unheard of, grief-stricken families were left to the mercy of private relief, although the manner in which Utahns rallied to the support of those families proved generous. More difficult to assess, however, were the causes of the explosion and the question of responsibility. Overall, the disaster and its aftermath sounded a dissonant note in Utah's coal fields and created repercussions that endured as the tragedy was perpetuated by stories, poems, ballads, and other folklore.[2]

Forty days before the explosion at Scofield, miners were forewarned of possible disaster. On March 22 a terrific blast caused extensive damage at the Pleasant Valley Coal Company's nearby Castle Gate mine.[3] Fortunately, the mine was empty at the time of the explosion. It was company policy in Castle Gate to evacuate the mine before the blasting powder was triggered by electricity. Such safeguards, however, were not practiced in the Winter Quarters mine.

At Scofield on May 1, many outside the mine thought the noise caused by the explosion resulted from someone setting off a blast in honor of Dewey Day.[4] A celebration had been planned for the children commemorating the second anniversary of Admiral Dewey's victory over the Spanish navy at Manila in the Philippines, and that evening a dance was to be held in the newly constructed Odd Fellows Hall.[5] Soon the residents of Winter Quarters and Scofield knew that something dreadful had occurred, and thoughts of Dewey and dancing were replaced with deep anxiety and later, for many, bitter remorse.

Within a few minutes of the explosion—which according to the stopped watch on one of the dead men occurred at 10:28 a.m.—a rescue party headed by the mine superintendent, T. J. Parmley, entered the mine. Several problems hampered the rescue effort. Afterdamp, a toxic gas mixture, remained in the mine, and two members of the rescue party were overcome by the lack of oxygen and were carried unconscious from the mine. Physically, the rescuers were nearly overcome by the stench of burned bodies, and emotionally they were stunned by the realization that the dead were friends and relatives. Mine officials had no record of who was in the mine at the fatal moment, and the rescue team feared that many men lay buried and the recovery of their bodies would be difficult.

The first rescue effort ended at 2:00 a.m. on May 2. After sixteen hours in the debris of the mine, the men were too exhausted to continue. The next morning miners from Clear Creek, Castle Gate, and Sunnyside arrived in Scofield. They were divided into parties and organized into shifts. By May 6 most of the bodies had been recovered. The first victim brought from the mine was Harry Betterson (supposed at the time to be John Kirton). Although burned beyond recognition, he was still conscious and crying out in agony for his comrades to end his misery. That night Betterson died.

Death for the miners came by one of three means: the explosion itself, falling coal and debris from the explosion, and the afterdamp or lack of oxygen resulting from the explosion. The men killed by the explosion were badly mutilated. Many corpses were found with the hands raised as if to protect the face. The eyes were closed and turned away from the hands indicating, perhaps, that the men saw the flash of the explosion.[6] These bodies were found in the rear of the mine and were the last ones to be taken out.

Those killed by the afterdamp were found in various positions: one man had filled his pipe and had sat down to light it; others still had their tools in their hands; one man was eating a sandwich; some men were in groups where they had apparently gathered to decide what to do. Other groups had been overtaken by the afterdamp as they tried to run ahead of it.[7]

In his report, State Mine Inspector Gomer Thomas said most of the men in the No. 1 mine—which connected to the No. 4 mine where the explosion took place—could have made it to safety had they run out of the mine when the blast occurred.[8] Most who attempted to flee tried to leave through the No. 4 mine because the way was shorter; instead of escaping the afterdamp they ran right into it.

The official number of dead as determined by company and state officials was 200—a figure far short of the estimates made by newspaper reporters, miners, and others who investigated the disaster. The miners who counted the bodies at the mouth of the tunnel listed 246 dead.[9] A week after the explosion, the Finnish community maintained that fifteen countrymen had not been recovered. Expressing concern about the number discrepancy, Governor Heber M. Wells wrote Mine Inspector Gomer Thomas on May 7 that "it is a most astonishing thing to me that there is so wide a difference of opinion in regard to the number of men killed."[10]

Remarkably, several men escaped death. The account of John Wilson is one of the most extraordinary tales to come from Scofield. Wilson, who was working at the mouth of the mine, was blown 820 feet down the canyon gulch; he suffered a crushed skull and was impaled through the stomach by a stick. Despite Wilson's critical injury, he recovered rapidly.[11] James Naylor was even more fortunate. He had been thrown 200 feet by the force of the explosion, but he was uninjured and able to aid in the first rescue attempts.

Roderick Davis managed to escape from the mine after the explosion and joined one of the rescue parties; but while inside the mine, he was overcome by the gas and fell unconscious. Taken for dead, he was thrown into one of the cars being filled with bodies. Once out of the mine, he was placed in a row of corpses to be washed. When the men began to wash him, he regained consciousness and walked out of the room.[12]

Approximately twenty of the victims were young boys who worked in the mine as couplers and trap boys, connecting coal cars and opening and closing doors inside the mine. However, many of their young companions had escaped unharmed. At the sound of the explosion, Thomas Pugh, a fifteen-year-old boy, seized his hat in his teeth and kept his nostrils covered while he ran a mile and a half to the entrance. He covered the entire distance in darkness without a light. Upon reaching the mine entrance, he fainted. His father, with whom he had been working, died inside the mine.[13]

The dead miners in the No. 1 mine were brought out in coal cars with as many as twelve corpses in a car; mutilated and burned bodies were brought in sacks. Once the dead were outside, clothing was removed, and soot and powderburns were cleaned from their faces. An identification tag with name and address was attached to each corpse, and the bodies were taken to the company boarding house at Winter Quarters. C. L. Nix, an employee of the Pleasant Valley Coal Company who issued coupon books to the miners, was sent by the company to take charge of the boarding house and to initially identify the dead as fast as they were brought out. After the bodies were again identified, this time by friends and relatives, they were washed and dressed in a white shirt and collar, a necktie, and a black suit, purchased by the Pleasant Valley Coal Company from ZCMI, the Mormon church department store in Salt Lake City. The dead who were members of the Mormon church were provided with the burial robes designated by that religion.[14]

Salt Lake City was able to furnish only 125 coffins, and 75 additional coffins were sent from Denver. After the bodies had been washed and dressed at the Winter Quarters boarding house, they were taken to the Scofield school. When the coffins arrived, the bodies were placed inside and then taken to the dead miners' homes. The fifty-one who were not to be buried in Scofield were taken by special trains to their destinations.

The 125 graves were placed in the northwest corner of the Scofield cemetery. The gravesites were merely trenches with coffins placed scarcely three feet apart. The remainder of the graves were dug in various family plots of the cemetery.

J. H. Eccles, the local sawmill operator, filled an order for 200 headboards. The inscriptions were made with a lead pencil. A *Salt Lake Tribune* reporter claimed that 50 percent of the names were misspelled and many had to be corrected by friends and relatives.[15]

On the day of the burial in Scofield, Thursday, May 5, a man stood at the gate of the cemetery and checked the names of those to be buried; another man matched each casket with the number of its assigned grave. A driver was then instructed where to take the casket. The weather suited the dismal occasion; it rained much of the day and at times a strong wind blew. The next morning a thick frost covered the ground, and higher in the mountains considerable snow had fallen.

Two burial services were held in Scofield. The first service, in memory of sixty-two Finnish miners, was conducted by Reverend A. Granholm, a Finnish Lutheran minister who had come from Rock Springs, Wyoming. The second service was performed under the direction of Mormon Apostles George Teasdale, Reed Smoot, and Heber J. Grant, and Seventies President Seymour B. Young. In his remarks, Reed Smoot counseled the survivors against rash words and acts toward the Pleasant Valley Coal Company. The survivors were admonished to make no demands of the company and to guard against outsiders or labor agitators coming in like vultures to "harrow up" feelings.[16] Burial services were also held in Salt Lake City, Ogden, Provo, Coalville, Springville, American Fork, Eureka, Richfield, Price, and other small towns in Utah.[17]

Because of the magnitude of the disaster and the necessary haste in taking care of the dead, some mistakes were made in identifying the bodies. Two caskets were exhumed in the search for the body of Thomas Padfield. On May 4 a joint committee of Odd Fellows and Knights of Pythias identified the remains of one of their members. The next day it was discovered that the body had been shipped somewhere in Utah and that the corpse of an unknown man was in its place. The mistake could not be corrected, and at least one family buried the wrong remains.[18]

There were stories of deep tragedy but none more touching than that of the Louma family. Seven sons and three grandsons of Abe Louma and his wife had left their home in Finland and had come to America, eventually arriving at the Winter Quarters mine. They wanted their father and mother, ages seventy and sixty-five, to live the rest of their days with them. The sons wrote their parents that they were earning more money than they had ever made in Finland and that it would be unnecessary for the old people to work any longer if they came to America. Abe Louma and his wife arrived in Scofield three months before the explosion. Six sons and three grandsons were killed in the disaster. Five of the sons and two of the grandsons were married. Only one son, Mataho Louma, survived.[19]

Newspaper reporters wrote melodramatic accounts reflecting the journalistic style of the day. Nevertheless, many of their reports serve as important historical documents of the Scofield tragedy. The following story from the *Salt Lake Tribune* tells of the indifferent manner in which mourners selected funeral attire.

> An incident now common, but always pathetic, is the selecting of funeral apparel at the company's store. The men who are left are working night and day in the relief shifts and at other duties, hence the matter of

> selecting devolves upon the heart-broken. The listless manner in which the poor women attend to this necessity is touching to the extreme. It only demonstrates again the paralized nature of their feelings, for they perform the function with a far-away cast of the eyes, accepting in almost every case what the clerks suggest.[20]

Expressions of sympathy and offers of help came from many people and in a variety of ways. Three railroad cars were loaded with flowers by citizens and school children from Salt Lake City and from other towns along the line to Scofield. Messages of condolence were sent by many, including President William McKinley. The tragedy received national and international attention. London's *Daily Telegraph* reported, "There will be deeper sympathy with America in this awful catastrophe than has been evoked by any event on the other side of the Atlantic since the loss of the *Maine*."[21]

Miners from Castle Gate, Clear Creek, and Sunnyside gave freely of their time in rescue efforts. Citizens of Scofield helped prepare the bodies for burial. When Mayor Earll feared he would not have enough men to dig graves, fifty Provo men volunteered to go to Scofield as gravediggers. Women went to Scofield to comfort the mourners and to help with the housework and care of the children. Bishop Lawrence Scanlan, the leading Catholic ecclesiastic in Utah, opened Saint Ann's Orphanage to all the fatherless children of Scofield. When a food shortage was reported in the town, Salt Lake bakeries donated 3,000 loaves of bread, and the Salt Lake Millers' Association sent 400 sacks of flour to be distributed to the victims' families. Other foodstuffs and clothes for the orphans were sent by people throughout the state.

An attempt was made to solicit financial help from the federal and state governments. In an editorial, the Logan newspaper, the *Tri-Weekly Journal,* pleaded that "the honor of the state demands immediate action," and called for Governor Heber M. Wells to convene a special session of the legislature to appropriate financial relief for the sufferers in Scofield. In Washington, D.C., Representative William H. King of Utah met with leading members of the House to discuss federal relief for the widows and orphans of the Scofield disaster. The representatives expressed sympathy for those stricken but, in contrast to later traditions of federal disaster relief for similar misfortunes, opposed federal aid because precedent was against such action.[22] With no help from either the state or federal governments, the burden of relief was left to the coal company and individual citizens.

According to one report, the Scofield funerals cost the Pleasant Valley Coal Company $25,000. Besides the burial clothes, the company provided the coffins, said to have cost $90 each. The company store erased an $8,000 debt that the dead miners had accumulated during the month of April. Company policy had been to sign out coupon books to the miners at anytime during the month; on payday the amount was deducted. The families received the full amount earned by the dead miners during the month of April without deductions. Instead of the usual line at the company offices on payday, mine officials distributed the pay privately to each home. The Pleasant Valley Coal Company also donated $500 to the family of each man killed, a total of $100,000.[23] For many Utahns of 1900, the company response seemed generous, though others saw its action motivated more by public pressure than by humanitarian feelings.

Donations also came from private groups and individuals. A central relief committee appointed by Governor Wells coordinated fund-raising efforts. In many Utah communities, local committees collected donations. For example, Thistle gave $100; Price, $500; Beaver, $400; Park City, $1,000; and Nephi, $500.[24] The miners at Castle Gate assessed every laborer and miner $2.50 to aid the Scofield widows and orphans. The Church of Jesus Christ of Latter-day Saints contributed $2,500 to the relief fund.[25]

Along with direct donations, fund-raising activities were undertaken for the relief effort. Numerous baseball games were played throughout the state; dances, concerts, theatrical performances, ice cream and cake sales, and even a children's "magic lantern" exhibition (which earned $1.06) were held. Of special interest was the publication of James W. Dilley's *History of the Scofield Mine Disaster*. This important document, which sold for $2.25, was issued in memory of the dead and as a fund-raising project. A total of $216,289.81 was donated by citizens and organizations throughout the United States for the relief of widows, orphans, and dependent parents.[26]

Victims and Blame

Not so charitable was the intolerance shown towards the Finnish miners in Scofield. Many branded them responsible for the ill-fated explosion. Mine Superintendent Parmley believed that prior to the explosion the Finns had secretly taken large quantities of powder into the mine to dislodge greater quantities of coal and thereby earn more money. When the giant powder was touched off, it ignited the dust inside the mine.[27]

In addition, the Finns were criticized for refusing to help in the rescue effort. Later, the attacks grew more severe. Under the headlines "Finns were Heartless" and "Disgusted with Finns," the *Salt Lake Tribune* accused them of not caring for their dead and of insisting that the company rather than the survivors should bring out the dead. Finally, nativistic sentiment characterized the Finns as "Vultures and Ghouls" when some tried to appropriate needed clothing and shoes belonging to the dead miners.[28]

Little is known about the Finns who found their way to Utah's Pleasant Valley and who died in the Winter Quarters disaster. Probably, they were recruited from the coal mines in western Wyoming, where many worked the mines near Rock Springs. Undoubtedly, they had made two or three stops before reaching Utah, and there is some indication they maintained contact with their countrymen in Wyoming and Minnesota. From what is known, most of Utah's Finns came from the west central coast of Finland, part of a larger exodus forced out because of population pressure, Russification policies, and the industrialization of agriculture. Of the sixty-two killed at Winter Quarters, more than a third came from townships just outside the port city of Kokkola. They were mostly young men, the oldest forty, the youngest fifteen.[29]

In Utah, the Finns vied with the Italians as Carbon County's first major ethnic group. The Finnish miners arrived in the 1880s and the Italians during the next decade. However, the Finns concentrated in Clear Creek and Winter Quarters, while the Italians settled first in Castle Gate and then later in Sunnyside. Unlike the Italians and Greeks who eventually made their way to Winter Quarters, the Finns did not leave their Pleasant Valley enclave. Perhaps the cooler mountain climate with its early snows was more attractive to these men of the North than the near-desert locations of the other coal mines. Perhaps they remained in Pleasant Valley because their ranks were not continually swelled by newcomers as were other immigrant groups. Finally their established Finn Halls and saunas were a haven from the ever-present hostilities of other immigrant groups. Perhaps because they were more concentrated and more isolated than other immigrants, they became easy targets for blame.

As the charges against the Finns mounted, others blamed the company for failing to keep coal dust levels at a safe minimum and to provide safe ventilation. Mine Inspector Gomer Thomas allegedly charged the coal company with negligence and with failure to sprinkle the mines properly. Later, Thomas denied having made the accusation. Some people felt state officials were partly responsible for not adopting and enforcing stricter safety laws.

Distilling the miners' attitudes from the cauldron of facts, rumor, and bias, a *Deseret News* reporter noted that some of the miners were "exceedingly bitter in their denunciation of the company and others . . . said it was one of those things over which no man has control and for which no man or men should be held responsible."[30]

Despite the complex accusations and denials, some evaluation can be made. Official investigations by a coroner's jury, the state coal mine inspector, and the state chemist all failed to find the company guilty in any way but concluded that the disaster was caused by an accidental explosion of black powder or an undetonated charge in the back of Winter Quarters Mine No. 4.[31] However, an information circular published by the United States Bureau of Mines, *Explosions in Utah Coal Mines*, noted that although the exact cause of the explosion had never been determined, it could have been prevented by removing dust, using water, and handling explosives properly.[32] The March 22 explosion in Castle Gate should have warned company officials and miners alike. Had the system of blasting only when the miners were not in the mines been used in Winter Quarters as it had been in Castle Gate, a tremendous loss of life could have been prevented.

On May 28, 1900, the Winter Quarter mines reopened. Some miners left Scofield because of the explosion, either moving on to other mines or finding new vocations. For those who remained in the mines and for the newcomers who took the places of the dead men, the disaster was a frightening and unwanted warning of what their own fate might be. The Scofield disaster was a monument to the sacrifice Utah coal miners had made and a warning to all who entered the mines of the offering they might have to make. The Scofield legacy served as a bitter reminder for the miners, fighting to secure and maintain a safer as well as a more equitable situation through strikes and union organization.

Chapter III

The 1901 Strike

The tragedy of the Winter Quarters mine disaster was a primary cause of Pleasant Valley labor unrest, which in 1901 erupted in an open confrontation between miners and the Pleasant Valley Coal Company when a strike was called.

The Winter Quarters disaster worked in two ways to promote labor unrest. First, it illustrated the high costs in human life and suffering demanded by the coal industry and the type of sacrifices coal miners were called to make for mine owners. Bishop Lawrence Scanlan, leader of Utah's Catholics, asked coal company owners to extend their efforts in every way possible to avoid another disaster, to consider the great sacrifices their employees made, and to "treat more justly, considerately and kindly those fellow men who . . . are obliged to work for them, and by whose labor, sweat and even blood they are being enriched."[1] It became apparent, as tensions leading to the strike developed, that the Pleasant Valley miners shared Scanlan's hope for better treatment by the coal company and articulated this sentiment throughout the strike.

Second, the disaster fostered the situation that led to the January strike by bringing new men to the mines. The 200 men who perished in the explosion had to be replaced. Moreover, many who had been spared in the May 1 tragedy concluded the risks of coal mining were too great and chose to end their mining careers while they still had a choice. Consequently, several hundred new miners had to be found to take the places of the dead and of those who, for whatever reason, did not return to the mines. The inexperienced new men were unable to produce the same quantity of coal; therefore, their wages were less than their predecessors.[2] With the recent death of 200 miners reminding them of a similar fate and with the new men suffering from aching muscles caused by long work and rewarded with insufficient pay, the miners sought a meeting with coal company officials to bargain for a more equitable arrangement.

Early in January 1901, a miners' committee was appointed to present a list of demands to W. G. Sharp, general manager of the Pleasant Valley Coal Company. The demands focused on two issues: pay increases and the appointment of checkweighmen to insure the miners were given proper credit for the coal they mined. The miners asked for a twenty-five-cents-a-day pay increase for all persons employed underground. This amounted to a 10 percent increase over the average $2.50 a day the miners earned. The company was asked to pay fifty-five cents per ton (seventeen cents higher than the going rate) for mined coal, which amounted to a 31 percent increase. Reflecting another facet of the growing breach between labor and management, miners later demanded three checkweighmen to represent them at the weighing scales. These checkweighmen were to be paid by deductions from the miners and were not to be considered coal company employees.[3] The miners justified their demands by pointing to prosperous times and high prices, and arguing that "we should have some share in the prosperity."[4]

The Pleasant Valley Coal Company responded quickly to the miners' demands through a letter written by General Manager Sharp and read to the miners assembled in the Scofield Odd Fellows Hall on January 13, 1901. Sharp countered that the Utah coal miners were earning as much in eight hours as the Wyoming coal miners earned in ten. Wage increases could not be granted because the large volume/low profit coal contracts the company had negotiated were based on the prevailing wage scale. The company agreed to collect money for paying the checkweighmen, but only from those miners who signed a paper authorizing the company to make the deductions.[5] Requiring miners to give written permission for deductions for the checkweighman effectively intimidated individual miners, promoted contention among the miners as a group, and insured that the company did not foster union attitudes.

The Scofield Odd Fellows Hall echoed with a lively discussion following the reading of Sharp's letter; but after considering their alternatives, the miners countered with a more modest proposal asking for an increase of ten cents a day instead of twenty-five cents and for an increase of ten cents per ton of mined coal instead of eighteen cents. The assembled men insisted on a prompt reply and announced if the demands were not met, they would strike.[6]

Sharp refused to reply, and for two tense hours the assembled miners debated a committee recommendation against striking. A vote was finally taken, and with a majority of only nineteen votes, those favoring the strike won 132 to 113.[7]

There was good reason for the close strike vote. Most of the miners had been at the mines only a short time and had not saved enough money for such a circumstance. In addition, local merchants were unable to support the strikers by extending credit; they had not recovered from the losses they absorbed when the debts of the Scofield disaster victims went unpaid.[8]

Those assembled in the Odd Fellows Hall also debated a proposal that water and timber men continue working in the mines in order to keep down dust and carry on other safety procedures. This was voted down. The day's business concluded with two committees being appointed to inform others that a strike had been called.

James D. Fowler, Peter Reed, and Robert Menzies were assigned to go to the coal camps in Clear Creek, Castle Gate, and Sunnyside, stating the grievances of the Winter Quarters miners and entreating their fellow workers to join with them. A twelve-man committee was appointed to meet incoming trains and to inform prospective workers that a strike was in effect and that they should either leave or refrain from working the mines. Later, a fifteen-man committee was elected to represent the miners and oversee affairs.[9]

Because of the strike's spontaneity, bordering on a wildcat walkout, the company was able to win some public support by maintaining that it had not been given time to consider the demands.[10] The miners countered that the company had been given sufficient time—six days.[11]

Newspapers suggested that opposition to the strike existed among the miners themselves for two reasons: first, because the strike involved only one mine, the company would merely increase its production at the other mines; and second, because the miners were not financially prepared to carry the contest, time became a significant factor. These elements, along with several others that developed, account for the failure of the strike, even though the strikers were able to shut down all the mines in Winter Quarters and Clear Creek.

As the strike progressed, the miners became more articulate in voicing their grievances and demands. In a letter directed to the general public on January 29, 1901, the strike committee outlined four justifications for the strike.[12] First, it felt demanding an increase in wages was within reason, because pay was so low miners could hardly clear expenses. In a somewhat contradictory vein, the strikers noted that only two families had applied for relief from the strike fund. Both had been in Scofield fifteen years, which "bears out our claim that the longer you stay in that place the poorer you get."[13]

A request for a semimonthly pay schedule instead of the usual monthly pay period was also made. As a second justification for the strike, the miners claimed that the coal company had raised the price of coal and that the miners were therefore entitled to a wage increase. The coal company denied that a price raise had taken place, but the miners obtained a report from the secretary of the interior to prove that there had been a price increase by the Pleasant Valley Coal Company.[14]

A third grievance centered on the unreasonable prices charged at the company stores. Miners in Scofield were more fortunate than Clear Creek miners who were forced to buy exclusively at the company store. The strikers charged that profits of 100 to 150 percent were made on such essential items as tools, oil, and powder.[15] Miners were obliged to purchase supplies for use in the mines. Those who lived in company houses or houses built on company property were under extreme pressure to purchase all supplies at the company store.

The fourth major issue was a strong belief that miners were being cheated at the coal scales. Officially, the company required the miners to load 2,240 pounds to get credit for a ton, the extra 240 pounds compensating for any rock or slate that might be in the coal. The miners insisted that they were forced to load 3,200 pounds for one ton and that the company was selling all of this as pure coal to the consumer. This contention had led to the original demand for a checkweighman paid by the miners.[16]

The first week of the strike was unusually quiet; but at midnight on Tuesday, January 22, men marched from Scofield to Clear Creek, six miles away and, that morning at seven o'clock, met with the Clear Creek miners. As a result of the meeting, the Pleasant Valley Coal Company received another ultimatum: if an agreement was not reached in Winter Quarters by Saturday, January 26, the Clear Creek miners would mount a sympathetic strike. They also added that if a settlement was reached, they would demand the same pay increase given the Winter Quarters men. The strikers, excited with this show of support, returned to Scofield, where a procession of 200 men and boys, led by a brass band, marched up and down the canyon.

On Sunday, January 27, after the company had failed to agree to a settlement, 400 strikers headed for Clear Creek, preceded by the American flag and the Scofield brass band. The school drum corps, led by J. W. Dilley, was assigned a place in the middle of the procession. They intended to meet with the miners who had promised to strike. After marching five miles, they met a committee of their own men who earlier in the day had gone to Clear Creek to attend a miners' meeting and who were returning to Scofield. A semicircle quickly formed around their sleigh. They reported that the Clear Creek miners had changed their minds and had voted not to go out on strike. Instead, miners in Clear Creek had given the coal company two days to consider allowing the State Board of Arbitration to settle the dispute. The Clear Creek miners would not work the next day, January 28, but instead they would hold a general meeting at 2:00 p.m. If the company rejected arbitration, a general strike would then be called for January 29.

At this point, with snow nearly four feet deep, a crude platform was erected under overhanging cliffs, among the giant pine trees. The president of the miners' organization, John C. Bell, mounted the stand and read a letter from the State Board of Arbitration offering its services in settling the dispute.

After discussing the letter, the men voted to accept the offer of arbitration if the coal company would do the same. Shortly after this, a group of Clear Creek men and the Clear Creek band were sighted coming around the mountain following a miner bearing an American flag. The now nearly 500 men, led by the two bands and the school drum corps, headed to Scofield where they paraded to the enthusiastic cheers and waves of the women and children before convening another meeting in the Odd Fellows Hall. Altogether it was an impressive, spontaneous demonstration.

At the meeting, a committee was appointed to talk with General Manager Sharp. As the committee filed out of the hall to carry out its assignment, the combined bands played "The Star-Spangled Banner."[17] The committee returned to report that Sharp had refused to discuss the situation with the miners and had said, "As the Winter Quarters miners are out on strike there is nothing to arbitrate."[18] Since the Clear Creek miners were not officially on strike, the general manager implied a willingness to submit their grievances to the State Board of Arbitration for consideration.

The miners, demonstrating their eagerness to settle the conflict, voted to resume work if the company would arbitrate and not discriminate against those who had participated in the strike. Once again the committee visited Sharp and forced him to reveal his true position—there would be no arbitration. After the miners received this news, another vote was taken; a motion passed to remain on strike until the company was ready to arbitrate. A committee of three—two men from Scofield, Lewis George and Pat Rooney, and one from Clear Creek, T. F. Hoye—was appointed to go to Salt Lake City and to present the case to the State Board of Arbitration. The meeting which lasted until one o'clock in the morning, ended with a rousing expression of thanks to Mr. and Mrs. William Edwards, who had opened their boarding house for those needing needing lodging (including some men who had been expelled from the company boarding house) and to John E. Inglis and Neil Matson, two Scofield merchants who had assisted the strikers.[19]

In Salt Lake City on January 31, the State Board of Arbitration composed of Colonel Enos A. Wall, John Nicholson, and J. S. Daveler, met in separate meetings with the miners' committee and with William G. Sharp. The miners still favored arbitration. Sharp, however, insisting he was meeting with the board as a private citizen and not as a representative of the Pleasant Valley Coal Company, told the board that the company would not arbitrate.[20]

The board could not force arbitration. It recommended that the miners return to work; perhaps then the company would submit the issue to arbitration. The miners, realizing that this would be futile, already were setting in motion plans which, if successful, might salvage victory.

One of the objectives of the strikers was to win the support and sanction of organized labor throughout the state. Following the meeting with the State Board of Arbitration, the committee gathered with representatives of the Building Trades Council in Salt Lake City.[21] T. F. Hoye and Lewis George presented the miners' position. Afterwards, a resolution was unanimously passed by the Building Trades Council, expressing full sympathy with the strikers and calling for a boycott of the Pleasant Valley Coal Company by all members represented by the council.

Two days later, February 4, 1901, a similar meeting was held with the Utah Federation of Labor (UFL).[22] After listening to the miners' committee describe the conditions that led to the strike, the UFL passed a resolution, again unanimously, denouncing the company as unfair and requesting all honorable men to cease trading with the Pleasant Valley Coal Company until it agreed to give just recognition to its employees.

Later, both the Building Trades Council and the UFL agreed to support the strikers by donating to their relief fund.

With the sanction and support of organized labor in Utah, the striking miners appeared to have enough influence and sympathy to obtain another important objective—spreading the strike to Castle Gate and Sunnyside. This, however, did not occur; failure to enlist the active support of miners in other coal camps of Carbon County became one of the primary reasons why the strike failed.

Newspaper reports proclaimed that the miners at Castle Gate and Sunnyside were content and would not strike.[23] The strikers, knowing this was the usual response of newspaper editors and coal company officials, ignored the reports and began an intensive campaign to bring the Castle Gate miners into the conflict.

At the beginning of the strike, a miners' committee from Pleasant Valley visited Sunnyside and Castle Gate to explain the reasons for striking. The committee assumed that the miners in these camps would not hesitate to join the strike. There were good causes for this assumption, at least regarding the Castle Gate miners. First, the Castle Gate mine was reputed to be the most difficult mine to work in the entire county.[24] And second, a large number of the miners were Italian; the popular consensus was that these "foreigners" were prone to strikes and radical activity.[25] In fact, in the mid–1890s, the Castle Gate miners had attempted their own strike but failed, partly because they were unable to win the support of miners from other camps.

However, before the Pleasant Valley strikers organized their Castle Gate campaign, company officials called a general meeting of the Castle Gate miners and denounced the strikers, and those who voted to boycott the Pleasant Valley Coal Company (which also operated the Castle Gate mine). In addition, other resolutions stated that many miners presently at Castle Gate had worked at the Winter Quarters and Clear Creek mines, earning three to five dollars a day, and that in their experience with the Pleasant Valley Coal Company, the men had "met with fair and honorable treatment in the adjustment of [their] grievances."[26]

The strikers refused to accept the resolutions as a valid representation of the Castle Gate miners' true position. They noted that the meeting had been presided over by Thomas L. Reese, an official of the Pleasant Valley Coal Company, and the mine superintendent had been instrumental in railroading through the resolutions.[27]

Realizing that gathering support from the Castle Gate miners was in jeopardy, strikers from Scofield went to Castle Gate. The morning after the resolutions were passed, approximately twenty strikers left Scofield by train for Castle Gate while another 125 men began on foot the thirty-mile trip through the snow-covered mountains. The marchers arrived in Castle Gate late that evening and spent the night huddled around the coke ovens in an attempt to keep warm.

The Scofield delegation was up early the next morning. As the Castle Gate miners made their way to the mines, the delegation tried to persuade them not to go to work. Unsuccessful in this effort and ordered off the company property, the strikers formed ranks, and led by a brass band, they began a parade along the public road. Premature and inaccurate reports of support by the Castle Gate miners were relayed by strike leaders to their anxious comrades in Scofield. However, that evening a group of Castle Gate miners, led by the mine superintendent and his subordinates, met at the coke ovens and voted not to strike. The strikers, still not willing to acknowledge these men as representative of all the miners, announced there would be a meeting at 10:00 a.m. the next day, February 4.

The strikers then traveled to Helper to find lodging for the night. Guards had been stationed at the Castle Gate coke ovens to prevent the strikers from spending another night on company property. In a letter signed by J. C. Bell, F. B. Walker, A. Sinden, and F. Douglas, the strikers recounted their mistreatment at Castle Gate.

> We had no place to lay our weary heads and limbs, for the company had already issued an order to close up all public houses in which we might be accommodated, and as all the dwellings and boarding houses are built on the company's property, the people of necessity had to abide by the order. We, therefore, had to parade the county road all night in the snow, for this was the only unforbidden ground upon which we could put a foot.
>
> We called a mass meeting that we had to hold out in the snow. We could not get the K. of P. [Knights of Pythias] hall, as it is built on the company's property. The law protecting and governing the same is that "no meetings or gatherings that are injurious or detrimental to the company shall at any time be held there." Thus the K. of P. cannot rent the hall to anyone only as the company says.
>
> We were therefore compelled to hold our meeting in the road and in the snow. There were 200 of the Castle Gate men who came to the meeting being led in the procession by the superintendent of the mine, bosses, and a squad of deputy marshals numbering about thirty. They would not

> listen to our grievances, but the superintendent of the mines made a motion to remain at work and not come out in sympathy with the strikers, which was seconded by the mine boss and voted down by the deputies and not a few of the company's pets that came down to the meeting with guns on their shoulders to drive us out of the town.
>
> The Scofield miners are treated more like dogs than men here at the present time. They are even forbidden to cross the track to the depot in order that they might send word to their fellows in Scofield and elsewhere. They will not even allow us to go near the company's store in order to buy overshoes to keep our sore feet dry and warm.[28]

The meeting scheduled for the morning of February 4 was postponed until seven o'clock that evening because most of the Castle Gate men were in the mines working. The evening meeting proved unsuccessful. During the meeting, word was received that the Clear Creek strikers had called off the strike and were going back to work. This was partially true. A group of Clear Creek miners had met in the company store, as the beleaguered miners were quick to point out, where an attempt had been made to end the strike. At this point, the company offered the strikers free transportation out of Castle Gate, which they refused.

A final meeting was held on February 10 in the Knights of Pythias Hall, the company finally granting strikers use of the hall if they would leave after the meeting. Speeches were made both for and against the strike. An Italian interpreted the proceedings for his non-English-speaking countrymen who, it was rumored, sympathized with the strikers' plea for support but recognized the futility of a strike. A vote was then taken; three-fourths of the Castle Gate miners voted to remain at work. Perhaps a few Italians voted in the minority, because that same week, the Castle Gate Italians took a major step toward a pro-union position with the incorporation on February 14, 1901, of a fraternal organization known as Societa Fratellanza di Mutuo Succorso, Principe Di Napoli, Loggia No. 77 della Federazione Colombiana della Societa Italiana D'America. This organization, founded primarily by southern Italians, was a counterpart to an earlier Italian lodge, Stella D'America, founded in Castle Gate on January 19, 1898, by northern Italians working in the coal mines.[29]

In the aftermath, twenty men who had spoken out strongly in support of the strikers were discharged, and one Castle Gate miner, Andrew Young, was threatened by a rifle-wielding guard for opening his house to the strikers.[30] The failure in Castle Gate ended the attempt to spread the strike to the other camps, and proved a major turning point in the course of the strike.[31]

After the outbreak of the strike, the Scofield and Clear Creek miners sought to strengthen their position by affiliating with the United Mine Workers of America. Accordingly, they sent a request to union headquarters in Indianapolis, Indiana, asking for an organizer. They were informed that an organizer in Columbus, Ohio, would come to Utah in the near future.[32]

Unlike later strikes, outside organizers played no role in the 1901 strike. On February 28, 1901, the *Eastern Utah Advocate* noted that the union representative from the East still had not arrived. A month later a short note in the same paper declared, "The Pennsylvania agitators have gone away and left us in peace."[33] If organizers did come, presumably they spent only a few days in the area because by the end of February the strikers had been defeated.

Mine owners, operators, and those opposed to the strike effectively used various methods to discredit the strike and the strikers. A frequently used argument was that the miners were lazy—they could make good wages if they were only willing to work. A *Salt Lake Tribune* correspondent interviewed several miners who maintained they could earn at least three dollars a day even in the most difficult sections of a mine. The correspondent concluded, "When a man makes less than $2.50 per day the trouble is in the man and not the mine."[34]

The morals of the strikers were attacked. The *Eastern Utah Advocate* stated that in Castle Gate the strikers were unable to get liquor at the local saloon because it had been closed and therefore were forced to rely on an Italian bootlegger who made a daily trip to Castle Gate from Helper with a barrel of wine.[35] It was later revealed that some of the men had gotten drunk in Castle Gate much to the disgrace of the strike effort.[36]

Rumors circulated that the old-time miners opposed the strike and that the strike leaders were recently arrived radical agitators, who "strangle[d] free expression on the part of the men opposed to the strike."[37] The *Salt Lake Tribune* went so far as to falsely blame the strike on the Italians in Castle Gate.[38] Others charged that the Finnish miners in Clear Creek and Winter Quarters had instigated the strike. An anonymous article in the *Deseret News* implied that the Mormon miners were not supporting the strike and that the trouble lay with the non-Mormon newcomers. The author noted that before the Winter Quarters disaster most miners were Mormon; and if they had a grievance, they took it to Thomas Parmley, who was both mine superintendent and the local Mormon bishop. The Mormon miners, the writer continued, were hard workers and were able quickly to reach an agreeable understanding with their mine superintendent/bishop. However, with the influx of non-Mormon miners after the disaster, Parmley's influence quickly declined, and the old structure, hewn from a homogeneous combination of economic, nationalistic, and religious components, was dissolved.[39]

John Potter, the Mormon bishop in Sunnyside and employee of the Utah Fuel Company, wrote to the *Deseret News* that, although they had had trouble in Sunnyside, company officials had always met with the miners and made adjustments so that the men were making good wages and wanted no trouble. In addressing his letter to the predominantly Mormon readership of the *Deseret News*, Potter made his position as a local religious leader clear. "There are a great number of our people here, and my advice to them as their Bishop is to keep out of strikes."[40]

This Mormon antilabor posture paralleled that of the Catholic church. Bishop Lawrence Scanlan, the highest Catholic official in the state, advised the Catholic miners at Clear Creek and Castle Gate to remain at work and not go on strike.[41]

Relations among the strikers were under continual strain as unending debate focused on the issue of returning to work. At one of the strikers' meetings the spokesman for the Finnish miners vowed that if the men returned to work without a settlement, the Finns would never go out on strike with them again. The band then played "I'll Not Go Out With Reilly Anymore."[42] Dissension within the strikers' ranks was punctuated by rumors that strikebreakers were en route from Colorado in response to a guaranteed wage of $2.50 a day.[43]

A primary strategy of coal operators was to breach the unity of the miners, and this was done in a number of ways. In Castle Gate, mine officials reminded the miners that at an earlier date, when they themselves had gone out on strike, Winter Quarters miners had ignored their pleas for help, and now was their chance to get even. In an attempt to frighten the strikers into believing their jobs were in jeopardy, company officials in Winter Quarters reported that men had applied to return to work. Also, antistrike observers pointed out to the striking miners that the committee they sent to Salt Lake City to meet with the State Board of Arbitration and with other union groups was comfortably housed in the Walker Hotel while the strikers were suffering in the coal camps.[44] With this constant barrage of antistrike sentiment, the strikers were unable to enlist broad public support. The antagonistic public attitude rang clearly but quietly in the collection plates of the miners sent to canvass the state for donations to the strike effort. The collection amounted to only $64.25.[45]

Although public support for the strike was minimal, it influenced the Utah State Legislature, in session at the time, to pass two pieces of legislation. First, because the State Board of Arbitration failed to act effectively in handling the strike, a bill repealed the 1898 statutes dealing with the Board of Labor, Conciliation, and Arbitration. This bill, which attempted to strengthen the board's powers, was introduced by Senator Orson F. Whitney of Salt Lake City and passed both houses of the legislature without a single dissenting vote. The 1901 law differed from the previous one basically in requiring that applications for arbitration be made prior to a strike or lockout and that applicants promise to abide by the decision of the board.[46] The second bill, introduced by Representative Frank H. Holzheimer of Eureka, prohibited companies from forcing, through "coercion, intimidation, threats, or undue influence," employees to trade at a company store or to live in a company-owned boarding house.[47]

Earlier, these very issues had caused strikes in Eureka in 1893 and 1900 and led to the election of Representative Holzheimer under the promise to curb the company's abuse in the Tintic Mining District. Their resolution also addressed some of the coal miners' grievances. A labor committee made up of representatives from the striking coal miners, the Utah Federation of Labor, and the Building Trades Committee were active in promoting the introduction and passage of these bills.

The almost complete lack of violence and lawlessness in the Utah coal fields was unique for the time. Despite this fact, many companies, anticipating trouble, hired large numbers of company guards for duty in the camps. Records indicate that about forty extra guards were hired in Salt Lake City to supplement an equal number of company guards already in the strike area.[48] The reason for the guards, according to the *Salt Lake Tribune*, was "the great amount of carousing which commenced when the strike commenced."[49] A miners' committee was organized to watch the guards, and only one serious incident took place between a company guard and a strike sympathizer. When a guard named Olsen went to McMillian's boarding house to pay his bill, McMillian and several others called him a scab and a blackleg. The agitated guard invited McMillian to step outside the house and go a round or two. McMillian obliged, and went for Olsen with a butcher knife. The armed guard pulled his gun and forced McMillian to return to the house. Later a warrant was sworn out by McMillian against Olsen for intent to kill.[50]

There were a few other minor incidents. In one, a Finnish striker unsuccessfully attempted to sabotage the powerhouse at the Winter Quarters mine by rolling a large boulder down the mountain onto the structure.[51] Another altercation (the only actual report of physical violence) involved Pat Wycherly, who was accused of hitting J. C. Davis at the entrance to the hall where a meeting of the strikers was to be held. Wycherly prevented Davis from attending the meeting because he thought Davis had jumped the strike and was working. The next day, assault charges against Wycherly were dropped when Davis, apparently guilty of working during the strike, was unable to get anyone to testify in his behalf.[52]

Despite the lack of violence, the strike, nevertheless, saw a suppression of the strikers' rights. Members of the strikers' fund-raising committee were quarantined in Scofield when an alleged smallpox epidemic threatened the town.[53] One striker, Joseph Caldwell, reported that while riding the train between Colton and Scofield he was forced at gunpoint to give up his seat to one of the company guards.[54] According to the strikers, Sheriff Hyrum Wilcox repeatedly threatened to "put the crowd of us into boxcars and ship us out of the country."[55]

After the strike was called, men were forced to leave the company boarding house. The leader of the strikers, John Bell, was notified by the company that his house would be torn down; the house had been built by Bell but was on company property.[56] While the strikers were in Castle Gate, the miners in that camp were threatened with instant dismissal if they gave food to the strikers.[57] When three Clear Creek strikers—Lewis Sholtz, John B. Doors, and John Lawrence, members of a band that had gone to Sunnyside to play for a dance—were discovered in a boxcar trying to make their way back to Clear Creek, the boxcar was sidetracked, and the men were locked inside without food or heat. Several local citizens protested until the men were taken to jail.[58] When the strikers in Clear Creek and Winter Quarters heard of the arrest, many volunteered to go to Price to force their release.[59]

Yet the bonds of unity were not strong among the strikers. Failing to win support of all miners in the county, the Winter Quarters and Clear Creek miners clearly had little hope of victory. There was much talk among the miners of quietly returning to work; although this idea was squelched when the Pleasant Valley Coal Company announced that returning miners would be forced to answer for their waywardness and to sign an "ironclad." The four stipulations of the yellow-dog contract required miners to swear they were not members of a union, that they would not join if an attempt were made to organize one, that they preferred to settle grievances directly with company officials rather than through union officers, and that they were satisfied with the Pleasant Valley Coal Company wages and methods of doing business.[60]

In response to the "ironclad," Lewis George, one of the more moderate strike leaders who had considered returning to work, resolved to continue the strike:

> Why should I or any other man be required to reveal to the officials of the Pleasant Valley Coal company the secret thoughts of the heart in regard to the personal welfare of myself or my fellowmen in order to secure a day's work? If the company hires me to do a day's work for it, and I do the work satisfactorily, is there any reason why I should not be paid for the work without having to tell them what my position or conditions were in the past, or what my intentions are as to the future? I think not, and I must say that I sincerely pity the man who does. To sign away my rights I will not and may God pity the man who will.[61]

Three days after Lewis George's statement appeared in the *Salt Lake Herald*, a final meeting was held in Scofield on February 20, 1901, and the strike called off. Many strikers were denied employment, and those fortunate enough to be reinstated mined coal at a reduced rate.[62] Admitting defeat, one of the strikers said: "The next time we go out the result will be different. We shall then have every miner in all the camps in the union and they will hang together better than this time."[63] Less than three years later, it seemed the prediction would be fulfilled.

Chapter IV

The 1903-4 Strike

The first years of the twentieth century were a time of tremendous growth for the labor movement in the United States.[1] In 1900, nationwide union membership was 868,500, and by 1904, it had more than doubled to 2,072,700.[2] Unionism moved forward on two fronts: an increase of membership in already existing unions and an increase in the number of new unions. Recent immigrants from southern and eastern Europe were the important elements in the burgeoning labor movement. In the coal fields of Carbon County, Finnish, Slavic, and Italian miners provided the strength behind Utah's most serious labor confrontation to that time, a confrontation that began in the fall of 1903 and lasted more than a year.

Frustrated in their past efforts to unionize, Utah miners found fresh hope in events outside the state. Following the successful anthracite strike of 1902 in Pennsylvania, the United Mine Workers of America conducted and extensive campaign to expand its control. Prime areas of concern were the far-western coal regions of Colorado, Wyoming, and Utah. Consequently, when a call for organizers came from the Carbon County strikers in 1903, the UMWA responded immediately.

The Utah strike coincided with a strike in the northern and southern coal fields of Colorado, but the Colorado strike proved both an asset and a liability to Utah strikers. Because of it, the national UMWA provided funds and organizers that otherwise might not have been available. The movement of organizers back and forth between Utah and Colorado provided ties between the two strikes. In addition, the ties were reinforced because some Utah miners, especially the Italians, had worked in the Colorado mines. Psychologically, all this helped sustain the Utah strikers. However, the Colorado strike soon developed into civil war, and many

Utahns feared that the Carbon County strike would follow the same course. Therefore, from the beginning, the strikers were denied public sympathy.

Although several emotionally charged issues clouded the Utah strike, the right of workers to organize irrevocably split labor and management. Because the Colorado coal strike began three days before the Utah strike, the Utah Fuel Company argued that Utah miners had no real grievances but only were striking in sympathy with their Colorado counterparts. The strikers denounced this assessment and put forth several specific grievances. They charged that they were commonly cheated at the weighing scales (miners were paid by the amount of coal they mined, not for the hours they worked), that they were taken advantage of through the scrip and company store system, and that coal operators frequently violated Utah's eight-hour workday law for miners. In addition to settling these grievances, the miners also demanded an increase in pay and recognition of the UMWA.[3] The Utah Fuel Company had offered the miners a 10 percent raise two days before the strike was called. This was to be accompanied by a 23 percent increase in the price of coal to dealers. The miners probably would have accepted the offer had the company been willing to recognize the UMWA. Realizing that concessions won without union recognition were temporary at best and that without union protection they most likely would lose their jobs, the strikers remained firm in their determination to achieve union recognition.

The Utah Fuel Company responded to the strike by announcing that the company store would operate strictly on a cash basis and that credit would be given to no one until the trouble was settled. The company then posted notices assuring employees who remained loyal every possible protection from interference or menace by the strikers. The company's chief detective, C. W. Shores, hired guards to insure enforcement of these policies. His recruiting methods reportedly were simple and straightforward. When an applicant arrived seeking employment as a guard, "Shores handed him a Winchester rifle, 'Let's see you load and unload that, before we can know whether or not you will suit.' "[4] Rumors circulated that if the strike were not quickly settled, people would "be going to bed in the middle of the day to keep warm."[5]

The Utah Fuel Company's public response to the strike appealed to nativistic sentiments. Company officials argued that support for the strike came from single men without families, those who had recently entered the coal fields, and the "worst element of foreigners." The old English-speaking miners, officials claimed, did not support the strike. The company continued to brand the walkout as simply a sympathy strike for the miners of Colorado and blamed labor organizers and agitators from that area for stirring up the previously contented miners.[6]

The company accusation that outside agitators had caused the strike lacked any basis in fact. The move for a strike at Sunnyside had been spearheaded by local leaders, who were supported, initially, by all ethnic groups of miners; ultimately, however, the strike received the most dedicated support from the Italian miners. Immediately after the vote to strike was taken on November 12, local leaders requested that organizers be sent to Sunnyside. The first attempt to bring in union organizers failed when a delegation on its way from Sunnyside to Price to meet incoming union officials was intercepted by a party of armed men (supposedly employed by the Utah Fuel Company) and was forced at gunpoint to return to Sunnyside. But this proved to be only a temporary setback for the union. By November 17 union leaders with the aid of the Italian organizer Charles DeMolli, who according to local legend secretly entered Sunnyside in a wagonload of hay, had succeeded in organizing a miners' union with an enrollment of 468 members.

At a mass meeting of strikers in Sunnyside, a vote was taken to determine whether to accept the company's concessions as a whole and go back to work. The vote demonstrated the Utah miners' commitment to the union; they rejected the company's concessions and proclaimed they would stay out on strike until the company recognized the UMWA. Organizers had convincingly argued that "it was impossible for an operator to be honest. The only way for them to be honest was for us to watch them, and we cannot watch them until they recognize us."[7] The next day only 33 of the 850 employees at the Sunyside mines reported for work.

One reason the strike began in Sunnyside was the influence of the Italian Fraternal Lodge Fratellanza Minatori organized in Sunnyside in 1902. The organization's constitution provided that if any member was found working where a strike had been declared for reasons of an "increase in wages or for whatever motive," he could be expelled from the lodge.[8]

The company reacted quickly and emphatically to the unionizing effort. On November 17, a regular payday at Sunnyside, all men not working were paid in full and discharged. Those occupying company houses were ordered to vacate. The company then launched an active recruiting program among farmers in the local areas to work the mines.

Enough strikebreakers were recruited to keep the Sunnyside mines operating, although production dropped.

Following the initial success at Sunnyside, the UMWA organizers turned their attention to the remaining Carbon County camps: Castle Gate, Clear Creek, and Scofield-Winter Quarters.[9] On November 19 the *Salt Lake Tribune* reported that the "Colorado agitators" had met with little success in Castle Gate. Yet by November 21, the Castle Gate mines were closed as a result of the strike—a feat the union was unable to duplicate in any of the other coal camps. On November 28, 627 men in the upper camps of Clear Creek and Scofield paid the $1.25 initiation fee and joined the UMWA.

With the closure of the Castle Gate mines and the spread of the strike throughout the entire county, the conflict entered a new phase: Carbon County Sheriff Hyrum Wilcox, fearing an eruption of violence, requested help from Governor Heber M. Wells. The governor responded by sending General John Q. Cannon of the Utah National Guard to investigate conditions in Carbon County.[10] Cannon found the coal camps in a state of excitement. Emotions were running high. It was a time for action. Union organizers were at the pinnacle of their success in recruiting members; approximately 1,200 miners had joined the union. Parades were held daily by the strikers, who, on occasion, masqueraded as capitalists, miners, and slaves. In Castle Gate, parades were led by the Italian brass band.[11] The articulate organizers effectively appealed to the miners' dignity as individuals and hammered away to forge greater class consciousness. Organizer William Price reported:

> I met the attorney general of Utah the other day and he said to me: "The men who dig coal in Utah make nearly as much, if not all as much, as the men who handle the pen." Now, in the name of God, are not the men who dig coal entitled to as much as men who whirl the pen? The attorney general told me he went into the homes of the miners in Scofield. "You didn't see any Brussel carpet there did you?" I said to him. Then I said: "Hasn't my wife as much right to wear a silk dress as your wife, Mr. Kramer's wife or Mr. Williams' wife." He said he supposed of course, if I could afford it, but I tell you, my friends, what these men say to us is: "Calico was made for you people and silk for us."[12]

The labor organizers continually urged their members to avoid any threats or acts of violence against those still working in the mines. They argued that only by peaceful means could public support for the strike be won. Nevertheless, threats were made and acts of intimidation were committed by a few of the strikers. The anti-union forces magnified these incidents, effectively swaying public opinion against the strikers. Some of the alleged acts included calling miners who were willing to work "scabs." James Polve was threatened with expulsion from the Italian lodge, Stella D'America, at Castle Gate if he continued to work.[13] Some men were prevented from going to work when their clothes were stolen or soaked with water. In Sunnyside strikers were accused of burning a partly loaded boxcar and of "murdering" a company guard's cow.[14] In Castle Gate and the Pleasant Valley camps, parading Italian strikers were denounced for showing disrespect toward the American flag by carrying it upside down. Union organizers countered that this was done out of ignorance, not disrespect. Also, threatening letters were sent to certain working miners. One of the most colorful notes was written to John Knudsen.

> You are hereby notified that if you scab here any more you had better be buying a coffin, and that pretty soon, too. And if you ride that rope tomorrow you will be hung tomorrow night. Now, you —— —— —— ——, you look out, because we mean just what we say. the men that rote this letter belong to the White Cap society, and if you work tomorrow you will see them tomorrow night.[15]

On the other hand, the situation was aggravated by the presence of numerous company guards who had been deputized to act in an official capacity for the county. They served without pay from the county but received five dollars a day plus room and board from the Utah Fuel Company. Some company guards sympathized with the strikers; however, the extremely high wages paid the guards—many of them farmers who sought winter jobs to supplement their income—compromised their sympathies.[16] Some guards intimidated and bullied strikers, trying to provoke them to lawlessness.[17] The strikers considered these guards "the scum of Utah—men who have no principles, who will sell their honor for gold."[18]

Upon General Cannon's return to Salt Lake City, Governor Wells met with state officials and members of the legislature. The group reached the unanimous decision, based on General Cannon's report, that sufficient cause existed for ordering state troops into the coal fields to preserve order.[19] The following day, November 23, 1903, Governor Wells called out the National Guard; and by November 24 the entire force—34 officers and 399 men—was ordered to Carbon County.[20] They took up positions in Castle Gate, Scofield, and Clear Creek. Approximately 300 actually went to Carbon County; some guardsmen who were university students, farmers, or ranchers were excused unless conditions deteriorated.[21]

According to newspaper reports and instructions given by National Guard officers, the troops were to help county officials protect company property from destruction and to prevent threats and acts of intimidation against miners who wished to work. Some guardsmen held this pro-company, anti-union attitude throughout the strike. However, many others, as a result of friendly relations with the strikers and growing disgust with the coal company, found their sympathies swayed.[22] J. L. Ewing, sergeant in the Nephi Company and editor of the *Nephi Record*, noted this shift in an editorial.

> When the militia was sent out it was supposed to be to protect the coal trust's property and the lives of its employees from a lot of anarchists and lawless people: but investigation so far, has shown that the strikers are quite as law abiding as are the emissaries of the coal trust and that the striking miners are just as much in need of protection from the hired guards of the trust, as the company property and men are in need of protection from the striking miners. . . .
>
> We have but little sympathy with the miners out there, and we have still less sympathy for the coal trust. In the first place if the trust is paying $3 to $8 a day to miners as they claim, or an average of $3.50 to $3.75 a day they have no need to hire Finns or Dagos. They can get the best miners in the country for such pay; and if they have imported under contract or otherwise a lot of anarchists, as they claim they are, from the old countries, they have brought their trouble upon themselves and ought to suffer for it. As time goes on it is becoming plainer and plainer that the companies have been trying to prejudice the public by coloring the reports that have gone from the coal camps to favor the companies and to grossly misrepresent the strikers.
>
> It is a very significant fact that most of the miltiamen who have returned are in sympathy with the strikers, and when they left for the coal camps the very reverse was true. . . .[23]

One of the National Guard's first duties was to assist Sheriff Wilcox in arresting organizer Charles DeMolli, whose background and personal magnetism made him a strong adversary. DeMolli was born in 1870 in Brussels, Belgium. His father was a horse trainer from Milan and his mother was a native of Como, Italy. Educated at the Institute of Milan, DeMolli served three years in the Italian army before going to work in the silk mills of Como. In 1895, according to DeMolli, he became involved in a strike and revolution in Como and was banished from Italy. DeMolli's account was contradicted by Italian consular reports that noted that he emigrated in 1895 after escaping from prison where he had been sent by a Naples court for three years on counterfeiting charges.[24] After coming to the United States, DeMolli began writing for Italian language newspapers in the East. He drifted to the coal mines of Pennsylvania where he renewed his interest in unionism. Later, DeMolli worked in the coal mines of Colorado until he became associated with the newspaper *Il Lavoratore Italiano* (*The Italian Laborer*), published at Trinidad, Colorado. The paper served as the organ of the United Mine Workers among the Italians. The *Salt Lake Herald* described DeMolli's charismatic appeal:

> DeMolli, the silver-tongued, whose influence with his fellow countrymen is so feared by the Utah Fuel Company officials . . . is in appearance far from being the wild-eyed anarchist he is pictured by his enemies. A tall handsome man in appearance, dressed in rough chinchilla jacket, flannel shirt, cordouroy trousers and laced boots of a miner, he has a handsome face, typically Italian, with a small slooping mustache. His voice is soft and his manners suave although at times he fires with enthusiasm over the subject he may be discussing . . . .
>
> Besides being big and strong and handsome, and with a magnetic personality, DeMolli is eloquent with tongue and pen in the Italian language. Not only this but he can talk in the native tongues with Finlanders, Slavs, French, or representatives of other nationalities. With his level head, shrewd judgment, college education, suave manner, and great magnetism, he is regarded as one of the strongest men affiliated with the United Mine Workers and he is idolized by his followers.[25]

This articulate Italian—harbinger of unionism and socialism—was a flamboyant and effective organizer in Utah's labor history. Moving among the Italian workers in the dual role of a John the Baptist and a Jesus Christ, preaching the dawn of a new order, temporal salvation, and individual dignity, DeMolli's name would be revered more than fifty years later by those young Italians grown old, who were touched by his personality and message.

On November 20 DeMolli and several other organizers went by train to Scofield, intending to organize the miners in Winter Quarters and Clear Creek. A committee of the Citizens Alliance and coal company employees met the organizers in Scofield and refused to let them detrain. DeMolli and his aides were told that it was useless for them to visit the camps because the miners had signed an agreement that they would not strike. DeMolli returned to Helper and enlisted the services of two cowboys who assured him that they would furnish an escort of twenty-five men if necessary and see to it that he arrived in Scofield without injury.[26] On November 23 the trip was made overland by horseback to evade company guards covering the railroads and highways. DeMolli's influence was felt immediately; within two days the strike had been organized in the northern fields, bringing the entire county into the dispute.

The success in Clear Creek and Winter Quarters was marred by DeMolli's arrest in Scofield on the morning of November 25. A detachment of guardsmen accompanied Sheriff Wilcox to the Lone Star Saloon where DeMolli was speaking to several hundred people. The strikers could easily have overpowered the detachment. Reflecting apprehension and fear, one guardsman nudged the soldier next to him and said: "God! This looks like this is about the end. If they rush us, we'll look like rag dolls laying around here."[27] However, when DeMolli saw the sheriff, he held up his hands and said in English: "Quiet down boys. Everything's all right."[28] As he left with the sheriff, DeMolli once again cautioned the strikers to avoid disturbances of any kind. Taken before the local justice of the peace, DeMolli was found guilty of disturbing the peace and sentenced to thirty days in the county jail in Price.

On November 30 a special court session in Price heard evidence in habeas corpus proceedings brought against the sheriff and jailer of Carbon County. After hearing the evidence, Judge Jacob Johnson quashed the proceedings.

An appeal was made for DeMolli to Justice William Burrows. The appeal and a change of venue to Clear Creek were granted. In Clear Creek, the trial lasted all day and late into the evening of December 4, but after ten minutes of deliberation, the four jurymen returned a verdict of not guilty. DeMolli was free.

Meanwhile, near the end of November, Dr. Giuseppe Cuneo, the Italian consul, arrived from Denver. Newspaper reports said he had come at the solicitation of the coal company. In Colorado, Cuneo had supported the Pinkerton Detective Agency in suppressing Italians in the Colorado coal fields by aiding in the search for possible fugitives from Italian justice. Charles DeMolli contemptuously referred to the consul as a "toothpuller, pork butcher, and sausage maker."[29] In any event, Cuneo quickly alienated himself from the striking Italians. First, they heard rumors that he had come to persuade them to go back to work. Further, Cuneo had not bothered to inform any of his countrymen of the visit, and upon his arrival at Scofield, he met with General Cannon and other officials before seeing any of the strikers. When Cuneo finally requested that leading Italians meet with him, some refused, and those who did agree to see him were almost prevented from doing so. Deputy marshals tried to break up a meeting at union headquarters called to elect representatives to speak with the consul; however, the deputy marshals retreated after being shown an order from General Cannon granting permission for the meeting.

Nothing came of the meeting. Responding to practical politics and perhaps to a personal philosophy and status different from his working-class countrymen, the consul managed only an expression of neutrality. In so doing, Dr. Cuneo failed the striking Italians of Carbon County. With his strong support, the strike might have acquired a legitimacy it lacked. Ultimately, Cuneo's failure to sustain the Italians was seen by many as evidence that the strike was unwarranted.

The attempt by the UMWA to organize the Carbon County coal fields did, however, acquire the support of the Utah Federation of Labor (UFL). This organization was composed of twenty separate unions that had been recognized in the state. On December 3 John Thal obtained a pass from Governor Wells to investigate conditions in the strike area and report them to the Utah Federation of Labor.[30] John Thal's report, accusing the coal company and its hired guards of unlawful acts against the strikers, was presented and adopted by the UFL on the evening of December 7, 1903.[31]

Before his arrest, DeMolli had wanted to meet with Governor Wells to discuss the strike. Following his acquittal, DeMolli traveled to Salt Lake City where he, John Thal of the Utah Federation of Labor, and two union attorneys, A. B. Edler and J. S. Fowler, met with the governor. DeMolli denied that the walkout had been called in sympathy with the Colorado strike and insisted that the miners merely wanted the company to allow union men to work alongside nonunion men. The company was still demanding that miners forswear any union affiliation by signing an iron-clad as a prerequisite for returning to work. Union officials not only failed to obtain any state support, but their cause was struck a severe blow when Governor Wells announced: "Mr. DeMolli, you are not welcome. We don't want you, and the sooner you are out of the state, the better we will be pleased. We will facilitate your going in every way possible."[32] The governor informed DeMolli that the strikers could not win the strike and that if they persisted in their efforts to obtain union recognition through a prolonged strike, public sympathy would turn against them to the point where "men from all over this state will shoulder their rifles and go down there and run the Italians out of the state."[33] The day after the conference, the *Salt Lake Herald* denounced DeMolli as an unfit representative or leader of an American union and stated that "every friend of the miners will hope that DeMolli will be discarded at once and invited to go elsewhere. If he isn't, the union might well prepare to lose its case with the general public."[34] Within a week, DeMolli had left the state, but not permanently. In February of the following year, he would return to Carbon County.

In the coal fields several isolated incidents threatened to ignite the explosive tinderbox of anger and resentment, but no serious violence occurred. While in Clear Creek, Sheriff Wilcox was forced by two shotgun-wielding Italian women to surrender his rifle. A Finn was arrested and charged with assault for throwing rocks at another Finn who wanted to work. On the road between Scofield and Winter Quarters, a fight took place when two Finns referred to John Burgess and Joe Ruff as "scabs"; with the arrival of a soldier, the Finns tried to run away but were caught and arrested. Reports of threats and intimidations against those who wanted to work were common. The peace of the area was further jeopardized by the arrival of a carload of wine for the Italian strikers on the same day the National Guard entered the northern camps. W. Lloyd Adams, a member of the Nephi Company, recalled:

> My first beat was between two large square houses filled with Italians. They had unloaded a great many fifty gallon barrels of wine during the day. It seemed to be a celebration in these houses as I walked my beat up and down between those houses sometimes women would shake their fists at me and talk in a foreign language and sometimes men. I wasn't more than 16 and of course I kept walking. I didn't rest a minute on that beat.[35]

Meanwhile, the Utah Fuel Company accelerated its efforts to curtail the activities of union organizers. Exerting great influence upon the civil authorities, the company saw to the arrest of most of the leading union officials. On December 8 William Marinelli, president of the Scofield union, and William R. Lewis, secretary, were arrested, convicted, and sentenced to twenty days in jail on charges of vagrancy. David Wilson, secretary of the Sunnyside union, was arrested on a charge of disturbing the peace. Most sensational, however, was the arrest of A. B. Edler, an effective, fiery attorney for the UMWA. A socialist, Edler had campaigned in 1900 for a position in the United States House of Representatives on the Socialist ticket. He had worked as an organizer for the

American Labor Union. Edler traveled the western lecture circuit as an orator under the title of "Champion of the New Civilization," urging his listeners "to make men free." Eugene V. Debs said to him: "You are a born orator who will some day be in the foremost ranks. The world needs more men like you."[36] When Edler arrived in Scofield on December 9 to defend the union men who had been arrested, he was served with a warrant for his own arrest on a charge of criminal libel. The charge grew out of comments that Edler had made to the newspapers about the company guards.

During the proceedings that followed, Edler was represented by attorney William H. King, a Mormon Democrat who had served in the House of Representatives and who later became a U.S. senator. King declared in one of his encounters with Utah Fuel Company attorney, Mark C. Braffet: "I have never seen a place where the desire to railroad a man through to jail was so manifest. . . . There is not another place where an attorney for a corporation is supreme to the court."[37]

At the preliminary hearing, Edler was released on bonds of $1,500. According to observers, the court session continually bordered on physical violence between the opposing lawyers. The Edler affair brought a strong warning from Governor Wells to the Utah Fuel Company officials, telling them to keep a tight rein on their guards. The governor also requested General Cannon investigate and submit a report on the Edler arrest and trial. After discussing the episode at some length, Cannon, in his reply to the governor, concluded:

> I am of the opinion . . . that the whole incident has been ridiculously magnified by Mr. Edler himself in the interest of his own notoriety, and by the "war correspondents," who in the absence of anything at all exciting in a campaign so entirely uneventful and peaceable as this, seize upon the merest pretext around which to weave a "story." The episode would not have warranted more than an inch of space in the ordinary newspaper column if occurring under normal conditions.[38]

The newspapers criticized the general's assessment of the Edler affair.

> Cannon has demonstrated his inability to observe and state facts without prejudice. If his sentiments and attitudes have been construed fairly he has no business in a position of judicial or semi judicial responsibility . . . he seems to assume that the state forces have been sent to Carbon County to be placed at the disposal of the Utah Fuel Company till further orders, the taxpayers to settle all the bills.[39]

The pressure upon union representatives eased for a few days; but on December 26, reports circulated that a warrant had been issued for the arrest of Con Kelliher and William Price, the most prominent organizers in the Carbon County fields during the absence of DeMolli. The charge was "Conspiracy against the preservation of law and order." Kelliher and Price left for Salt Lake City when they learned of the warrant. However, upon returning to Carbon County, they were not arrested. Later, a warrant was issued for the arrest of Sam King, an attorney for the union and a major in the National Guard. The younger brother of William H. King, Major King was a judge advocate in the National Guard and served as a watchdog of guard activities against the union. In 1922, when his brother was serving in the Senate, Sam King would represent the strikers of that decade. The charge against King was "trespass" and had been sworn to by a company guard. Upon learning that King planned to remain only a few hours in Scofield, officials did not serve the warrant. Louis Roberti, who succeeded Charles DeMolli as union organizer among the Italians, reported that company guards arrested him on a vagrancy charge. When Roberti said the name was wrong on the warrant, company guards scratched out the original and substituted his name.[40]

Tension continued to mount in the coal camps. Sunday evening, December 20, a public confrontation in Sunnyside resulted in the arrest of thirty-three Italians. While at a dance, Virginio Bonvicini, suspected of spying for the Utah Fuel Company, was asked if he would prefer going back to work rather than moving into a tent. He replied that he would prefer returning to work. At this point, six or seven men began calling him a "scab" and other offensive names. When Bonvicini reported the matter to Justice George Hill, the Italians were arrested and charged with disturbing the peace. Nineteen of the thirty-three were found guilty and were fined. Two men paid, but seventeen served time in the county jail in default of payment. They were described as being a "jolly good natured set and accepted their sentences of imprisonment laughingly."[41] Each Italian wore in his cap a feather plucked from a stuffed eagle in the Knights of Pythias Hall in Sunnyside where they had been confined under guard.

As the strike continued, expressions of antiforeign sentiments grew more frequent. The coal company continually blamed the Italians for the strike, claiming that it was the foreigners, not the English-speaking miners, who were aroused by the labor agitators. The *Salt Lake Herald* echoed the nativistic refrain. "These threats have not come from American citizens, but from an alien class, ignorant in most cases, unlettered, and led by individuals who have little regard for the law and less for the institutions of civil liberty."[42] The editors also noted that although there was no better citizen than the Italian from the north, there could be no more undesirable citizen than the southern Italian. However, among the Italians themselves, the usual antagonisms, sentiments, and traditions that generally divided northern and southern Italians were suppressed; the old-world animosities were set aside for the good of the new-world union.

As part of the antistrike rhetoric, Italians were sometimes characterized as being opportunistic and niggardly. For example, James A. Harrison, Carbon County commissioner, commented: "The Italians as a class, live on about one-third what it costs the American miner, while ... they make the same amount of money for the same work. They have money saved to leave the country or remain in idleness for months. . . .[43] The *Deseret News*, reflecting a similar point of view, charged, "These Italians have refused to amalgamate with Americans or learn the English language and have lived with the intention of getting out of this country all they could and then returning to their native land of olives and dirt."[44] When Adjutant General Charles S. Burton visited the coal camps, local citizens told him that even though the Italians were posing as the most peaceful and inoffensive people possible, they stood with a dagger concealed, ready to plunge it into someone's back.[45] In a move to use this nativistic prejudice to its advantage, the Utah Fuel Company announced that if it could hire enough English-speaking miners, the foreign-speaking would never again be permitted in its mines. Coal officials lamented that it had been necessary to hire foreigners in the first place because most American-born miners would work only in fall and winter, leaving early in spring to return to their farms.

Discontent with the situation in Carbon County spread to members of the National Guard, many of whom found their experience unpleasant for a variety of reasons. For the first twenty days, the common soldier was paid $1.50 a day from state monies; but beginning with the twenty-first day of active duty, the federal pay scale of forty-three cents a day was in force. This caused much discontent because many of the men had left good-paying jobs with no assurance their positions would be held for them. Additionally, temperatures during November and December were low, and the inexperienced guardsmen suffered accordingly; at least one private on guard duty was overcome by cold and was found unconscious at his post.[46] Company H in Castle Gate became infested with lice and had to be fumigated.[47] With little to do execpt carry out the monotonous task of guard duty, the soldiers also suffered from boredom. One sergeant from Salt Lake City had written home so often describing the inactivity of the troops with the phrase "war is hell" that he began receiving letters addressed to "War is Hell" Tobias.[48]

As weather permitted, the guardsmen pitched horseshoes and played baseball and football. In Clear Creek, the Nephi troops borrowed sleighs from the local Finnish children to use on the snow-covered hills. Occasionally, dances were held by the guardsmen where the only way to tell the "males" from the "females" was the former wore hats and the latter did not. But idleness remained a problem and created strained relations among the soldiers. At one point, two Nephi soldiers, after a disagreement, got their guns and started for the woods. One of them was heard to say, "We're going up into that timber and one of us isn't coming back."[49]

For a time, some soldiers tried their hand at mining in the employment of the coal company, but this lasted only a few days. Certain guardsmen became jealous of others working for pay. After some soldiers barely escaped a cave-in, the company decided that hiring soldiers might create an unpleasant situation, and it no longer sought their employment.[50]

Although the guardsmen had been enthusiastically welcomed by the local coal company, strikers received them with mixed feelings. In Clear Creek, the Finnish community offered the soldiers the use of its social and cultural facility, Finn Hall, for their barracks. But in Winter Quarters, the Finnish miners denied the soldiers use of a similar hall. In a practical sense, the strikers were glad the National Guard was there, in the words of Charles DeMolli, "to protect them from the company guards and sheriffs." From an ideological standpoint, however, the strikers disliked the presence of soldiers because it implied a state of lawlessness—a condition attributed to the strikers, not the company.

Despite newspaper reports and statements by government officials to the contrary, the public was not convinced that the Guard was necessary. From the day the Guard was called out, friends and relatives tried to persuade officers that certain soldiers were not needed or that conditions warranted their remaining at home. A clever appeal was sent to General Cannon by a man from Provo on behalf of his son who had recently married. With scriptural authority to uphold the request, the sincere father recounted the admonition in Deut. 24:5, "When a man hath taken a new wife, he shall not go out to war, neither shall he be charged with any business; but he shall be free at home one year, and shall cheer up the wife he has taken."[51]

Many citizens resented the large expenditures necessary to maintain the Guard in the coal fields. Others were skeptical of the Guard's purpose in Carbon County, especially after the Utah Fuel Company had said it would continue to discriminate against union men.[52] Though it would be thirty years before the issue was settled, many were beginning to ask why state taxes should be used to deny citizens their right to join a labor organization.

With the cold weather, inactivity, and approach of Christmas, many guardsmen were anxious to return home. The discontent was summarized in the following letter written by a soldier in the field.

> We have seen no trouble and there has been evidence of none. I understand public sentiment is against the whole thing and it ought to be. We wouldn't any of us kick if there were need for us, but as it is this stunt will finish the national guard. I will ask for a discharge as soon as I reach Salt Lake, and almost all the company will do the same.[53]

Only two problems prevented the return of the militia before Christmas: the reopening of the mine at Castle Gate and the eviction of strikers from company houses, scheduled for the first of January.

The reopening of the Castle Gate mine was significant for several reasons. First, the mine had been the only one shut down by the strike. In addition, it was a hard place to work, and there was some concern that enough miners could be found to permit reopening. On December 11, 1903, representatives of the strikers met in Helper and voted unanimously to return to work, provided the company did not discriminate against union men. The following day, David Crow, mine superintendent at Castle Gate, posted a notice that the mine would reopen on December 21 and that applications from strikers and other miners would be accepted until noon on December 15. At first, the strikers were jubilant about the coal company's apparent concession to hire union members. However, company officials demanded that prospective employees forswear the union; they argued that those miners who had remained loyal to the company would refuse to work if union men were hired. The strikers, on the other hand, charged that the company wanted to break the union completely. As a result, the call to apply for work was ignored by the union miners.

At another meeting held in Helper on December 18, union representatives drafted a formal address to C. W. Kramer of the Utah Fuel

Company that listed their grievances and at the same time expressed their desire to return to work provided no discrimination was made against union members. They waited in vain for a reply. In Castle Gate the union had played its last hand. Now it could only wait to see if the newly recruited men would be capable of taking over the strikers' jobs.

Since the first days of the strike, recruiters for the company had scoured the entire state looking for men to work the mines. The recruiters were particularly successful among the Mormon farmers of Emery and Sanpete counties where church teachings against secret organizations (as the UMWA was branded), nativistic fears of the foreign-born miner, and economic concerns worked to favor strikebreaking efforts. Superintendent Sharp acknowledged the strong support of Mormon leaders in the surrounding valleys and noted that "many farmers now at work in our mines came here not only because they needed work, but because they realized that if the company lost this strike their farms would become practically worthless."[54]

At 5:00 a.m., on December 21, the Castle Gate whistle called the sleeping strikebreakers to work. Deadlines had been drawn at the east and west ends of the camp, and groups of six company guards patrolled each end. On a hill overlooking the camp, guardsmen with their Gatling guns in place were ready for any trouble. Estimates of the labor force recruited throughout the state ranged from the union calculation of 25 to the company figure of 100. Perhaps the company figure included would-be strikebreakers who instead fleeced the company by verbally accepting the jobs, along with clothing and coupon books from the company store, and then escaped with the booty before loading one shovelful of coal. Regardless, the company succeeded in reopening the mine.

Strikers were evicted from company houses and removed from company property primarily without violence. The one exception was in Winter Quarters on December 27 when twenty guards evicted the unmarried Finnish strikers. The guards had also ordered three married Finnish strikers to move and had attempted to force them out of their homes, though the married strikers had been given until the first of January to leave. The guards were attacked by the Finns, and several guards were severely beaten. The fate of the guards might have been worse had not one of their number escaped and run to the National Guard for assistance. Company F of Manti hurried to the scene. The company guards ordered the soldiers to open fire on the mob, but the order was ignored. Through a simple request, the soldiers were able to persuade the Finns to cease hostilities. Company F had been the Christmas guests of the Finnish strikers the night before and were recognized as friends.

After their eviction, strikers from various company towns gathered in such labor union camps as Mitchellville (apparently named for John Mitchell, president of the UMWA) located on property near Sunnyside belonging to George C. Whitmore, a long-time antagonist of the fuel company. Near Helper, the strikers gathered at the Half-Way House, a square one-story rock structure similar in style to the wood frame houses of the coal camp. This house and adjacent small farm in the mouth of Price Canyon between Helper and Castle Gate were owned by Paul and Caterina Pessetto, immigrants from northern Italy. Here, and at other places, the strikers lived in tents provided by the union. Some men felt that the newly recruited labor force would soon prove inadequate and that the company would be forced to take them back. Others, less optimistic, left after a short stay. They were encouraged in this by the coal company who arranged with the D&RGW to give the strikers a reduced rate—one cent a mile—on one-way tickets out of Carbon County.

The company faced a particularly ticklish problem with the houses, 225 of them were built by strikers on company land. Legally, the houses belonged to the coal company, but public sentiment would not tolerate the company evicting strikers from houses they had built. A compromise was negotiated between G. W. Kramer of the Utah Fuel Company and William H. King, attorney for the strikers. A committee of three appraisers was appointed: one by the company, one by the union, and one by the two appointed appraisers. The committee would appraise the value of each house and the improvements made by each striker. It would also fix a reasonable rental value on the property. The strikers then would leave the houses but would be paid the equivalent of six-months rent determined in advance by the appraisers. The strikers also would have the right to sell the houses before October 31, 1904, subject to company approval. All of the strikers, with the exception of a group of Italians in Castle Gate, agreed to the compromise.[55] The Italians balked at this arrangement for good reason. Only three or four people were ever reimbursed despite the careful work of the appraisal committee, which calculated a total value of $60,000 for the strikers' houses and improvements.

Following the uneventful reopening of the mine in Castle Gate and the relatively peaceful eviction of the striking miners, almost everyone realized that the National Guard was no longer needed.[56] In retrospect, many citizens commented that the calling out of the entire National Guard had been unnecessary and that the same duty could have been performed by a corporal's guard in each of the coal camps. By the first week of January, more than half of the soldiers in Carbon County had been relieved, and by January 24, all of the troops had returned home.

The Utah Fuel Company continued its recruiting attempts throughout the state; and as a result, an issue that had been brewing finally surfaced: since most of the strikebreakers were Mormon, the church's attitude toward labor and unionism suddenly demanded definition. Early in the strike, a *Deseret News* editorial "advised coal miners who want to work . . . at good wages, to respond to the call for decent, English speaking men by the Utah Fuel Company."[57] A month later, when Angus Cannon —president of the Mormon Salt Lake Stake, uncle of General John Q. Cannon, and himself an owner of mining claims in the Dugway and Mercur mining districts—announced in the Salt Lake Tabernacle that employment in the coal fields of Carbon County was available to anyone who wanted it, the issue burst into the open. Con Kelliher, organizer for the UMWA, reacted to Cannon's announcement by charging that "the Mormon church had commenced a fight to annihilate union labor in Utah." He appealed to organized labor throughout the United States to unseat Mormon Apostle Reed Smoot from the Senate who, because of his position in the Mormon church, "would be a dangerous foe to labor in Congress."[58]

After Reed Smoot's election to the United States Senate in 1903 as a Republican, several Salt Lake City clergymen and others felt that Smoot's high position as an apostle in the Mormon church would restrict his ability to perform senatorial duties free of church influence and control, and they protested his seating. The detractors felt he could not obey the oath to uphold the United States Constitution, and some falsely accused Smoot of being a polygamist. A subcommittee investigation recommended that Smoot be excluded from the Senate. Kelliher sent letters to Samuel Gompers, president of the AFL, and John Mitchell, president of the UMWA, requesting they use their power to prevent the seating of Smoot. Kelliher reasoned that Smoot in the Senate would be a dangerous and powerful enemy of labor and that his unseating would serve to rebuke the Mormon church for its intervention in the Carbon County labor dispute. However, President Theodore Roosevelt and other Republican leaders, who saw Smoot as a valuable asset to the party, interceded, and Smoot was allowed to take his seat in 1906.

Locally, the strikers were convinced the Mormon church acted in league with the coal companies to prevent the success of the UMWA in Utah. David Wilson, financial secretary for UMWA Local No. 2630 in Sunnyside, reported that the announcement in the Salt Lake Tabernacle and in other Mormon meeting houses of available work in the coal fields had resulted in a large number of farmers requesting work in the mines.[59] A miner, Thomas Phelps, declared: "Our Mormon people are raising scabs from the cradle. They have preached to them to go to the mines to take our places. They are under the impression that if the union comes to Utah, the corporations will close the mines."[60]

This furor prompted a telegram from the LDS First Presidency to Gompers and Mitchell defining the official church position.

> Reports have been circulated that the Mormon Church had used its influence against organized labor, and had endeavored to fill the places of union strikers with non-union men, we hereby emphatically deny that the Church over which we preside has taken any such steps or issued any instructions on this matter. The whole story originated in a false newspaper report.
>
> Joseph F. Smith, John R. Winder and Anthon H. Lund.[61]

If fears of adverse publicity—of the kind spawned by the Smoot hearings—prompted Joseph F. Smith and his counselors to shun an open admission of involvement, Mormon church records indicate that they nevertheless had instructed local church authorities to support the company's strikebreaking activities. On December 2, 1903, H. G. Williams, president of the Utah Fuel Company, wrote to President Joseph F. Smith explaining that Bishop John Potter of Sunnyside had sent Brigham Gould to Emery County to recruit men for the Utah Fuel Company. Gould's efforts were nullified by Jack Coombs and David Wilson, two striking miners, who followed the company representative and argued that he was misrepresenting actual conditions. Williams requested that President Smith reassure the people of Emery County about the good intentions of Bishop Potter and Brigham Gould.[62] The request was honored. In a letter sent to Reuben G. Miller, president of the Carbon Stake, the First Presidency clearly supported the company's strikebreaking efforts. Miller was instructed to telephone or write the bishops of his stake at once to assure them that the statements of the Utah Fuel Company and is representatives could be relied upon and that those who wished to work "would do well to pay no attention whatever to the statements of Coombs and Wilson."[63]

The strikebreaking effort succeeded. In a confidential letter to H. G. Williams, the First Presidency congratulated the Utah Fuel Company on its victory and offered advice on how to prevent further trouble. Williams was counseled to hire sufficient guards to protect the company's property and "keep the lawless men under control." In order not to be "obnoxious to the strikers," the guards were to be "discreet men" who were not to make arrests outside company property. The letter explained that this new advice was being offered because "you are aware how anxious we were that you should not be forced to yield to the demands, inspired by outside agitators, which the strikers made upon you, and we are still anxious that the victory won may be complete."[64] In the future, Mormon anti-union sentiment would plague Utah's coal miners though church leaders would become much more discreet in their pronouncements and actions.

Because Mormon miners did not remain in sufficient numbers, the company was forced, despite public promises to the contrary, to fall back on "foreign labor." Initially, a few blacks and Japanese were hired. Later twenty-five Greeks arrived to work in the mines, thus beginning the great influx of that important ethnic group to Carbon County.[65] In time, the "new foreigners" would lose faith in their employers and turn to the UMWA for redress of their grievances.

Meanwhile, during the first part of February 1904, Charles DeMolli, who had left the state under pressure from the governor in December 1903, returned to Carbon County. While in Sunnyside, he and Joseph Barboglio, local treasurer for the UMWA who later founded the Helper State Bank and became one of Utah's most respected businessmen, were arrested by company guards when they tried to cross a deadline on their way from Sunnyside to Helper. They were taken to the city jail and held without charges being preferred. That evening, when company guards arrived to take the prisoners to the company boarding house for supper, they found that the men had escaped by taking off the bolts on the inside of the door. DeMolli and Barboglio were caught late that night below the coal camp, taken before the local justice, and found guilty of breaking jail. They were sentenced to thirty days in jail and fined thirty dollars. After an appeal both were released on a $250 bond.

A week later John Trewarther, president of the Scofield miners' union, and Thomas Casley, secretary, were arrested, tried, and found guilty of conspiring to interfere with trade and commerce by persuading men to strike against the Pleasant Valley Coal Company, to shut down the mines, and to make it impossible for the coal company to fill orders for coal.[66] Bonds were placed at $1,000, which Trewarther and Casley were unable to obtain. Three weeks later their attorney succeeded in having the bonds reduced to $500, and the men were released.

Strategically, the arrest of union leaders continued to be an effective tool to crush the strike. Although legal justification for the arrests and convictions was highly questionable, the policy as voiced by one company guard seemed to be "damn the legality, as long as we make it stick."[67] The arrest of union officials was insignificant, however, when compared to the arrest of 120 Italian strikers following the arrival of Mother Jones in Carbon County.

Mother Mary Jones was one of the most forceful and colorful leaders in the history of American labor. Born in 1830 in Cork, Ireland, she immigrated to America as a child. In 1861 she married, but six years later, while living in Memphis, Tennessee, she lost her husband and four small children in a yellow-fever epidemic. Following this tragedy, she went to Chicago where she engaged in dressmaking before her recruitment as a frontline lieutenant in America's industrial conflict. To workers, Mother Jones represented the best qualities of Joan of Arc and Mother Mary. But to her enemies, she was a "vulgar, heartless vicious creature, with a fiery temper and a coldblooded heart brutally rare even in the slums."[68] She was accused of being a well-known character in the red-light districts of Omaha, Kansas City, Chicago, and San Francisco. In Denver, she allegedly had her own brothel, and the *Deseret News* reported that she was a good friend of the notorious Kate Flint, one of the pioneer prostitutes of Salt Lake City.[69] But vice was not the seventy-three-year-old, white-haired woman's stock in trade. Though she appeared to be a feeble grandmother, she stood as a dedicated organizer whose sole purpose was to improve the workers' condition through effective and recognized union affiliation. Many years earlier, she had cast her lot with organized labor; and with the single-mindedness of a partisan with nothing to lose, she fought with equal zeal for both the native- and foreign-born miner.

Mother Jones was active in the Colorado coal strike until April 1904 when she came to Utah. Her arrival in Helper characterized her flamboyant style.

> She walked to the best hotel and announced who she was. Later she entered the dining room and sat down to dinner. At the close of her meal the waitress placed a finger bowl before her. "Take it away my girl," she said in a voice that could be heard all over the dining room,

> as was doubtlessly intended, "such things are not for me, they only give some poor overworked girl extra work at washing dishes."
>
> . . . This sentiment was applauded as being as noble a one as ever uttered by the white haired old woman dressed in black. The words were passed from mouth to mouth and the railroad hands who boarded at the hotel were filled with admiration for her sterling qualities.[70]

The Utah strikers met Mother Jones at a gathering on April 17, the first Sunday after her arrival in Helper. Following the meeting, she and several Italians visited William Price, an UMWA organizer who had been active since the first days of the strike in Utah. Price was in quarantine with smallpox and officials attempted to put Mother Jones, because of her exposure to the disease, in quarantine for the customary fifteen days. A shed had been constructed in Helper for those exposed, and it was intended that Mother Jones spend the fifteen days there. Just before her consignment, however, two Italian strikers, Celeste Dalpiaz and Angelo Pilatti, set fire to the shed damaging it so severely it could not be used.[71] Mother Jones refused to remain in voluntary quarantine; and on the morning of April 21, Deputy Sheriff Harry World moved to arrest her, and Sylvester Tedesco, and Caesar Antonia, for violating quarantine regulations. World later testified that upon his arrival at the Half-Way House, where Mother Jones was staying, he was prevented from making the arrests by at least 100 armed Italians who occupied positions at the house and the rocks and prominent points surrounding the camp. Deputy World and county officials then requested that Governor Wells send the National Guard back to Carbon County to prevent possible violence and to see that the arrests were made.

Governor Wells again asked General John Q. Cannon to inspect conditions in Carbon County to see if the situation warranted recalling the Guard. Once more emotions ran high. Rumors circulated that Mother Jones planned to lead the strikers in a march to Castle Gate in an attempt to regain possession of the houses built or occupied by the strikers before eviction by the coal company. Company guards in Castle Gate reported that they had been fired upon by Italian strikers and that some thirty shots had been exchanged. When the strikers persuaded Italian coal heavers working for the D&RGW to leave their jobs, the strike seemed suddenly to be gaining momentum.

After hearing Cannon's report and despite the pleas by county officials, Governor Wells decided against sending the National Guard back into Carbon County. The governor could not forget that the $25,000 he had been forced to borrow from a New York bank to finance the Guard's earlier activities in Carbon County had become a serious blemish on his administration's record. Criticized for the first mobilization of the Guard, Wells hesitated to order a repeat performance.

Following Governor Wells's refusal to send the National Guard, Sheriff Wilcox began organizing a posse to go to the Half-Way House to arrest those guilty of violating quarantine regulations. The sheriff recruited approximately forty-five men, and at 2:00 a.m. Sunday, April 24, the posse left Castle Gate for the Half-Way House. Concealing themselves in the rocks above the camp, the posse waited for daylight. At dawn, they charged down the hill and within a short time had taken 120 Italians into custody. In her autobiography, Mother Jones gives the following account of the raid:

> On Saturday night I got tipped off by the postoffice master that the militia was going to raid the little tent colony in the early morning. I called the miners to me and asked them if they had guns. Sure, they had guns. They were western men, men of the mountains. I told them to go bury them between the boulders; deputies were coming to take them away from them. I did not tell them that there was to be a raid for I did not want any bloodshed. Better to submit to arrest.
>
> Between 4:30 and 5 o'clock in the morning I heard the tramp of feet on the road. I looked out of my smallpox window and saw about forty-five deputies. They descended upon the sleeping tent colony, dragged the miners out of their beds. They did not allow them to put on their clothing. The miners begged to be allowed to put on their clothes, for at that early hour the mountain range is the coldest. Shaking with cold, followed by the shrieks and wails of their wives, and children, beaten along the road by guns, they were driven like cattle to Helper. In the evening they were packed in a box car and run down to Price, the county seat and put in jail.
>
> Not one law had these miners broken. The pitiful screams of the women and children would have penetrated Heaven. Their tears melted the heart of the Mother of Sorrows. Their crime was that they had struck against the power of gold.[72]

The *Deseret News*, in a more restrained manner, also expressed sympathy for the Italians.

> When light enough the sheriff gave the order to charge the camp, compel every man to come out of his tent and put him under arrest. The

> orders were carried out. Many were hustled out of bed in their night clothes but were allowed to return and dress. The women also were forced out while the tents were searched for firearms. Some guns and ammunition were found. . . .
>
> When the officers were taking the strikers away the women were very abusive in their language and some threats were made. The men were very peaceable and made no resistance. As the strikers were being marched to the train, their wives and children who had followed them down to Helper, were weeping as they said good-bye. The scene will not be forgotten soon by those who witnessed it.[73]

Since the county jail in Price was far too small to house 120 prisoners, the Italians were sent to a "bullpen" on property owned by E. M. Olson where they were housed in a shed that had been used to store carriages. The entire area was surrounded by a high fence patrolled by guards.[74] Local residents and hotels were asked to help by cooking meat for the prisoners. The Mormon bishop, Ernest S. Horsley, gathered food for them, and water was obtained from private cisterns or hauled from the Price River. Sheriff Wilcox claimed the Italians were better off in the bullpen than they had been at home, but others argued that the prisoners suffered because of inadequate shelter.[75] Although the situation seemed quiet, there was some portent of violence. One prisoner was transferred to the jail after he threatened to burn down the shed. Rumors circulated that an attempt would be made to free the prisoners. Attorney Mark F. Braffett of the coal company called the prisoners "desperate marauders." Italian women, who came with food and concern for their men, badgered the officers and guards in broken English that "put to shame a drunken pirate."[76]

On the morning of April 27, the 120 Italian prisoners were loaded into boxcars of a special train and were taken to the schoolhouse in Helper to appear before Justice M. A. Ward. The charges covered a multitude of lawless acts, including resisting an officer, rioting, intimidating men and driving them from work, and disturbing the peace. Attorneys for the Italians pleaded not guilty in each case. The next morning the court began trying each of the strikers individually. The outlook was that the trials would drag on for several months.

By May 13, eleven Italians had been found guilty and an equal number had been discharged. At this time, the Italians remaining in the Price bullpen were also discharged. Several factors led to the dismissals: the wish to avoid lengthy court proceedings, insufficient evidence, the high cost to the county of feeding the prisoners, and a good deal of public sentiment against the arrest and imprisonment of seemingly peaceful and innocent Italians.[77]

During the first week of May, while the trials of the strikers were underway in Helper, Mother Jones left Carbon County and traveled to Salt Lake City. Two important conventions were being held in the state's capital: a convention of the Western Federation of Miners (WFM) and a convention of the Utah Federation of Labor. The Utah Federation had issued a call for all unions throughout the state to send representatives to its convention being held in conjunction with the Western Federation of Miners. Following the defeat of the UMWA, it was apparent to the other unions that their future efforts might be crushed just as easily. This was the reason for the attempt to develop a strong union alliance within the state. Mother Jones attended both conventions and, because of her fame as an organizer, was invited to play an important role in the meetings. Given a standing ovation by the delegates and asked to speak, she praised the downtrodden miners of Utah and Colorado and graphically described the abuses of the coal companies and government officials.

Mother Jones found the conventions an excellent forum for expanding and promoting her ideas of socialism. The Utah Federation of Labor was facing the choice of joining with a national organization. The moderate delegates tried to affiliate the Utah Federation with the AFL. The more radical groups, including the WFM, advocated joining the socialists. The vote not to join the moderate AFL but to remain independent was something of a victory for Utah socialists. After the convention, Mother Jones left Utah and went to Cripple Creek, Colorado, to continue her crusade for the laboring class.

From the time the governor was asked to send the National Guard back to Carbon County until the end of the court proceedings against the Italians, a vocal segment of the public had protested the actions against the strikers. The unofficial leader of this group was the ambitious and imaginative Sam Gilson. Prior to his arrival in Price, Gilson had led an exciting life. At the age of fifteen, he had left his home in Illinois and had traveled to Nevada, where he entered the livestock industry and supplied horses to the Pony Express. As a United States marshal, Gilson supervised the execution of John D. Lee—a polygamist Mormon—for his role in the 1857 massacre of California-bound emigrants at Mountain Meadows in southwestern Utah. Later Gilson discovered and promoted many mining properties and invented several successful machines, including a hydraulic ore concentrator and separator. The mineral gilsonite took its name from him.[78] This black-maned giant of a man and notorious Mormon-baiter was a crusty frontiersman who drifted ahead of the developing frontier. As a businessman, he was deeply involved with the Saint Louis promoters of gilsonite and was assumed to be as callous to the strikers' cause as anyone. Yet, as in all other situations, he was a maverick choosing to throw his energetic support to the underdog coal miner.

In reply to Deputy World's request for National Guard troops to assist with those who were violating quarantine regulations, Gilson, in a letter to the *Salt Lake Herald*, criticized the sheriff and his deputies for their methods of serving quarantine notices and asserted that everything the local law officials "have done has been in the interest of the Utah Fuel Company and not for the peace and quiet of the county."[79] On the day following the arrest of the 120 strikers, Gilson was arrested on a charge of abusive language when he allegedly swore at Sheriff Wilcox and Deputy World. Gilson was released on a twenty-five-dollar bond, and during the next few weeks, he held several "indignation meetings" to protest the actions of the coal company and county officials.

While the Utah strike dragged on, union support gradually eroded. During the first week of the imprisonment of the Italians in the Price bull-pen, the executive board of the UMWA adopted a motion instructing union officials to close up affairs in District 15, which included Colorado and Utah. John Mitchell, president of the UMWA, in admitting defeat, noted that while the strike in District 15 had been the most expensive in the union's history, the mines were being operated successfully. Mitchell declared that it was the union's intent to end the strike as soon as possible but not to abandon the work of organization in Colorado and Utah.[80]

On May 26, UMWA representative Harry Bousfield addressed a letter to H. G. Williams asking for a meeting to discuss the strike. Williams had taken over affairs for the coal company after the death of G. W. Kramer in Denver early in May 1904. Bousfield's letter was an attempt to implement the motion passed by the union's executive council and to obtain permission for the miners still on strike to return to work. In reply, Williams declined to meet with the union representative and parroted Kramer and other company officials, stating that the Utah Fuel Company had no dispute with its employees, but if it did, it would gladly meet to resolve any differences. Instead, according to Williams, industrial peace, tranquility, and prosperity reigned in the Utah Fuel camps; and if there were a threat to this situation, it would come from "parties not now in the employ of the company and in no way connected with it and to whom the company had no employment to give."[81] Realizing that the union was anxious to call off the strike, the company determined to make capitulation as humiliating as possible.

By late July, the union had ceased direct financial support of the strikers, and several men turned to rustling local cattle and sheep for food. The Utah Fuel Company became more zealous in its efforts to rid the area of strikers. David Wilson, the hot-blooded secretary of the Sunnyside local who had been arrested earlier in the strike for disturbing the peace, wrote to the *United Mine Workers Journal*:

> The Utah Fuel Company deputized about twenty armed guards and marched down to what is known as Strike Town and rounded up all the strikers upon a trumped up charge and they then marched them to what is known as the dead line and there the strikers were warned by M. P. Braffet, solicitor for the Utah Fuel Company that if they did not leave the country within two days they [the company] would organize "a band of lawabiding citizens" better known as scabs, who would come down to Unionville and put a rope around every man's neck who had not left the country and string them up. They, the strikers, were also required to sign a paper which they were not allowed to read, and when they remonstrated they were politely informed that if they did not sign their names they would have their heads cracked . . .
>
> Some of the papers in this State often made the assertion that the great coal miners' strike in Carbon county is a thing of the past. Then in the following issue they contradict themselves by publishing transactions that occur in the coal fields. If any one attempted to enter the several coal camps of Carbon county he would realize whether the strike was a thing of the past or not, for if he were a stranger the first obstacle he would encounter would be rifles in the hands of company guards, and then he would be escorted to the office of the superintendent and if unable to give a good account of himself he would be railroaded out of town with the admonition that if he returned more stringent measures would be taken in his case.[82]

During the last week of November 1904, $7,000 was sent to the First National Bank of Price by the UMWA. The money was to be used for railroad fares and other necessities in helping strikers leave Carbon County. The funds were administered by James D. Ritchey and R. M. Smith, officers of the union. They visited the strikers and encouraged them to accept the offer and to seek employment elsewhere. Although some strikers complied and were sent to the mines of Colorado, many declined the offer and were left to their own resources; the union refused to provide further aid.

Abandoning the strike brought strong criticism from many Utah strikers who had taken the promises of continual union support literally. John Mitchell was charged with selling out to the coal companies. One Italian striker named Marchiori at Sunnyside bitterly described the UMWA as "putrid and corrupt" and stated that the union had brought the Utah miners only ruin.[83]

In January 1905, the Scofield Local No. 188 of the UMWA wrote Robert Randall, a delegate from Dietz, Wyoming, to the 1905 Annual Convention of the UMWA. The letter strongly attacked Mitchell and his handling of the Utah strike and requested that Randall inform other delegates of Mitchell's misconduct. At the convention, Randall was questioned by the credentials committee about his remarks in Dietz, Wyoming, denouncing Mitchell as a traitor to the union movement. The challenge infuriated Randall, and he requested permission to report about the strike in Utah and Colorado. In his speech, Randall strongly criticized Mitchell and, as part of his remarks, read a letter signed by John Macketa of the Scofield local.

> John Mitchell has been a traitor against us. John Mitchell told us at the beginning of the strike of Carbon county that he would support us for three years if necessary, but we found out that he is a traitor, and also he said he would stay with us; but no, he went to Europe in the month of June, 1904, and he sold himself to the company, and in the first part of December he sent a man over here to try and get us away from here by trying to give the fare to us. He has been telling us to get support whereever we can, and that he would settle all the bills, and the bills were about $10,000; but no. . . . If Chas. DeMolli was the president of the U.M.W. of A. we would have won the strike long ago. John Mitchell and his officers have arrested Chas. DeMolli because he knew we would win the strike if Chas. DeMolli was free. John Mitchell is only a bluffer and not a union man . . . he is only a traitor.[84]

The convention responded by defending Mitchell and by passing a resolution demanding that Randall publicly retract his charges. Randall refused and left the convention. The attack by Randall, despite its courageous nature, served no purpose except to console those Utah miners who were convinced that Mitchell had betrayed them. For John Mitchell, the Scofield letter undoubtedly dampened any interest in future organizational efforts in Utah.

The remaining strikers, most of whom were Italians, did not return to the mines. Several became farmers along the Price River and raised produce for the miners who had taken their places; others went into business in Helper. Although the D&RGW was the parent company of the Utah Fuel Company, many Italians found work on its railroad gangs. Some continued to hold to the promise of unionism and turned to the Western Federation of Miners for help.

The Carbon County miners who stood with the UMWA during the strike suffered a humiliating defeat, but they were not alone. Ironically, the working miners faced reduced wages in January 1905.

From today's perspective, the demands made by the union in 1903 were clearly justified. The Utah Fuel Company was willing to negotiate all demands except union recognition. To the miners, however, this was the most important item. As a result, the struggle became a fight for union recognition with the question of working conditions de-emphasized. This gave an advantage to the coal company, because many regarded the union as a foreign import that, with its radical tendency toward socialism, had no place in democratic America. Given the prevailing fear of radicalism, the strike failed, even though the public had no particular sympathy for the coal company.

Company forces made the most of the public's nativistic feelings, urging that the strike was led and supported by foreigners who could not and would not allow themselves to be assimilated into American society. Thus one's patriotism was challenged if one supported the union. Many saw the dispute as a struggle between a ruthless company, which was at least American, and a foreign-dominated labor union.

In the beginning, the company convinced the public that a state of lawlessness existed in the coal fields and that the strikers were to blame. Later, it was apparent that the company guards were as guilty, if not more so, of violence and intimidation as were the strikers. However, in light of the confrontation between strikers and guards a remarkable characteristic of the strike was that there was relatively little violence. This was due in large measure to the belief of union officials that the blame for any violence would be charged to them. Therefore, they continually urged strikers to conduct themselves peacefully. The presence of the National Guard in the coal fields served as another deterrent. Although the top officers of the Guard were procompany, the men themselves maintained a fairly neutral position throughout the strike. In time many guardsmen became good friends with the strikers, although they did not sympathize with the union; some even reacted strongly against the company.

Governor Wells proceeded with caution, but he exerted more pressure on the union than on the coal company. He demonstrated a marked, but not atypical, prejudice against the Italians and other foreign-born miners. His public contempt for Charles DeMolli appeared on more than one occasion, and he backed off from his attempt to settle the strike when the Utah Fuel Company refused to arbitrate. Wells also failed to act against the use of the courts by the company to restrict the activities of union officials. Company guards who were causing serious problems received better treatment than union members whose offenses were minor. No company employees or supporters were arrested for their activities. Also the rights to free assembly and free speech were denied union officials. However, much of the situation was out of the governor's control and in the hands of local officials, who were naturally cautious about attacking a company largely responsible for the area's economy and for paying 78 percent of the local taxes. During a period of prosperity, it was not as easy to find fault with the company as it would be during the depressed thirties.

For its part, the union was strengthened by the nationwide efforts to unionize the coal fields and by the ability of the organizers: Charles DeMolli, Con Kelliher, and William Price. At the same time, the UMWA succeeded in signing about two-thirds of the miners; yet it failed to maintain the loyalty of many who joined; and they soon left the area. The continuing support of the foreign strikers proved a liability in obtaining public sympathy for the union.

In the end, the strike failed because the union simply did not have enough support, either internally or externally, to win against a powerful and influential company that effectively played on reactionary antiforeign sentiments in defending its position.

Chapter V

Intermission for the United Mine Workers of America, 1905-17

The controversial end and bitter aftermath of the 1903–4 strike angered Utah coal miners about the role of the UMWA. This hostility and the union's preoccupation with other issues halted all UMWA activity in Utah until 1918. Nevertheless, worker unrest continued in the Utah coal fields as the Western Federation of Miners (WFM) attempted to salvage the 1903–4 strike for organized labor, and as newly arrived Greek immigrants molded themselves into a new type of Utah coal miner.

The WFM was organized in 1893 in Butte, Montana, by forty delegates from western metal-mining camps. The catalyst for the organization was the 1892 Coeur d'Alene, Idaho, strike where strikers, opposed to a proposed wage reduction, battled company guards. Major strikes occupied most of the early history of the WFM, including strikes in Cripple Creek, Colorado, 1894; Leadville, Colorado, 1896; Coeur d'Alene, Idaho, 1899; Telluride, Colorado, 1901; and Cripple Creek, Colorado, 1903.

Initially part of the American Federation of Labor, the WFM left that organization in 1896, charging that Samuel Gompers, AFL president, concentrated on the East and that the AFL was reluctant to raise large sums of money to aid the WFM with its strikes. In 1898, the WFM sought to promote the interest of all western unionists by organizing the Western Labor Union in Salt Lake City as a counterpart to the AFL. The WFM desired to emphasize industrial unions in contrast to the AFL's perceived emphasis on craft unions. Despite repeated pleas by the AFL for reaffiliation, WFM leaders felt that the gulf separating the two organizations was too great, and in 1905, the WFM played the leading role in organizing the Industrial Workers of the World, considered by historians as America's most radical union.

Although the interests and objectives of the WFM and the UMWA were similar, the UMWA and its president, John Mitchell, remained an active and vital part of the AFL. This philosophical and political difference led WFM leaders to criticize the conservatism of UMWA leaders and their handling of the Utah coal miners' strike. As a militant union, the WFM was willing to step in where the less committed had failed. The Utah coal miners' situation provided the challenge and opportunity. Even before the 1903 strike began in Utah, Charles Moyer, president of the WFM, remarked, "John Mitchell need not be surprised should his membership turn their faces to the West, seeking affiliation with organizations which have sent out their message to the world, that as labor produces all the wealth, such wealth belongs to the producers thereof."[1]

In March 1905, in a letter to the editor of the *Eastern Utah Advocate*, Louis Viljiquist announced, "The Western Federation of Miners Union has taken the Carbon County strike under its care and all the strikers had joined the Western Federation."[2] Viljiquist's report was somewhat overstated, and at the 1905 annual convention of the WFM, held the last week of May in Salt Lake City, President Charles Moyer reported on the Carbon County situation. He noted that in February 1905, he received a charter application from 100 coal miners in Utah. In their letter of application, the miners explained that they had been members of the UMWA and, in response to the union strike call, had endured great suffering, including losing the homes they had built on company ground. With the strike lost, John Mitchell was now asking them to return to work. This would be impossible unless they disbanded their union. After an investigation of the Carbon County situation, the executive board of the WFM voted unanimously to charter the coal miners and to provide whatever assistance the treasury would allow. Moyer, sensitive to the plight of the Utah strikers, maintained, "If there was ever a body of men worthy and entitled, not only to the consideration of the Western Federation, but to organized labor in general, it is the striking coal miners of Utah, deserted by the organization of their craft."[3] However, Moyer was quick to review the recent struggles of the WFM, and although offering moral support, he cautioned that the WFM would be unable to provide financial assistance to continue the strike.

Undaunted in its primary desire to maintain its union organization, the Carbon County group pursued its request, and in mid–March 1905 a charter was granted. The WFM rendered aid through its chief attorney, John H. Murphy, who reportedly was sent to Utah to assist the miners legally regain their homes. In addition, from March until August 1905, the WFM sent regular monthly payments totaling $1,100 to Celeste Dalpiaz, financial secretary of the local union.[4]

The WFM was active in the Carbon County coal fields between 1905 and 1911. Two locals, No. 237 in Helper and No. 204 in Scofield, were organized. The Helper local, also known as the Castle Gate local, lasted from 1905 to 1911; membership in Local No. 237 numbered 80 in 1908, 179 in 1909, but dropped to 24 in 1910.

Records indicate the Scofield local existed only from 1907 to 1909.[5] In 1911, the union was disbanded, and the WFM's No. 237 was assigned to the Dearing Smeltermen, which organized on May 8, 1911.[6]

The WFM also took up the cause of Charles DeMolli, UMWA organizer in Utah and publisher of *Il Lavoratore Italiano.* After his return to Colorado from Utah early in 1904, DeMolli resumed his work with the Italian newspaper. In one of the issues, DeMolli responded to the attacks of a female anti-union reporter from Denver, known as "Polly Pry," by branding her a shameless prostitute. DeMolli was arrested and convicted for sending obscene materials through the mail. The UMWA refused to support an appeal, and newspaper reports claimed the radical DeMolli had lost favor with John Mitchell. Speculation flourished that Mitchell had initiated the charges against DeMolli in an effort to get rid of him.[7] Abandoned by the UMWA, DeMolli found hope when the WFM appointed its chief counsel, John H. Murphy, to conduct his appeal. The effort proved unsuccessful, and DeMolli served a two-year prison sentence in the Ft. Leavenworth Federal Prison.[8] After his release, he worked for a time as an organizer for the WFM in California.[9]

Turmoil marked the eastern Utah coal fields for several years after the 1903–4 strike. In March 1905, the *Eastern Utah Advocate* reported that the Carbon County farmers, who had entered the mines as strikebreakers, were returning to their homes. "Few, if any, are a cent ahead . . . and many are losers because their property . . . has suffered during their absence."[10]

In November 1905, a fire completely destroyed the tipple and crushing plant of the Utah Fuel Company at Sunnyside. The $160,000 fire was initially blamed on saboteurs, but later the cause was identified as a coal dust explosion that ignited the timbers of the tipple.[11] Regardless of the cause, Utah strikers no doubt rejoiced in the company's misfortune.

During 1907, several events indicated the unrest in the coal camps. In March recently imported Greek and Austrian (South Slavic) miners at Winter Quarters left the mines because they had been cheated by the coal company at the weighing scales. Discontent reached such proportions that Leonidas Skliris, labor agent for the Greeks, journeyed from Salt Lake City to Winter Quarters to negotiate with the Greek miners and coal company officials.[12]

Stirred by company injustices, disaffected miners attacked employees of the Utah Fuel Company. In Sunnyside, Hyrum Wilcox, Carbon County sheriff during the 1903–4 strike and in 1907 a deputy employed by Utah Fuel, suffered a fractured skull from a rock thrown by Joe Charron, a Sunnyside miner, during a Fourth-of-July celebration. The *Eastern Utah Advocate* reported that this was but one of numerous assaults on Utah Fuel employees throughout the county.[13]

In Kenilworth, 100 miners struck for three days the last week of October, protesting the Independent Coal and Coke Company's attempt to add a third shift to help push development of the newly opened mine. The miners argued that the men selected for the third shift "would have practically nothing to do but draw their pay." The other two shifts would be required to do all the blasting and loading.[14] The coal company, anxious to meet production schedules, did not challenge the miners but reportedly settled the problem to the satisfaction of the strikers.

The Union Pacific mine near Winter Quarters, which had been shut down in 1897, was reopened in 1907, and immediately labor problems developed. Initially, the miners were paid less than those employed in the nearby Utah Fuel mines at Winter Quarters and Clear Creek. Their wages were raised to equal those of the Utah Fuel miners, but some dissatisfied miners called for organizers from the UMWA to help secure a more favorable rate.[15] Apparently the UMWA organizers did not arrive, and the Union Pacific miners turned to the WFM for support. The *Eastern Utah Advocate* claimed that within the Scofield, Winter Quarters, and Clear Creek area there were more than 500 members of the WFM and that the union was gaining ground as the Union Pacific miners joined its ranks. In late December 1907, the article noted, 323 members had attended a union meeting in Scofield.[16]

Finally, on January 4, 1908, forty-four men were discharged from the Union Pacific mine. The official explanation pointed to declining demands for coal; however, union supporters argued that the layoffs were due to the miners' union affiliation. Despite the *Eastern Utah Advocate* report on the strength of the WFM in the Pleasant Valley area, there is no evidence the union aided the Union Pacific miners. Agreeing with the coal company position, a report by WFM officials concluded that the problems at the Union Pacific mine were primarily economic and not related to the union question.[17]

The efforts of the WFM in the Utah coal fields were, perhaps, even less successful than those of the UMWA. The local unions were never strong enough to strike against the coal companies or even to affect the rehiring of miners such as the forty-four men dismissed at the Union Pacific mine in January 1908. However, through the WFM, some financial help was made available to the unemployed strikers after the UMWA called off the 1903–4 strike. The manner in which the WFM offered to assist the defeated Utah strikers encouraged the coal miners and accentuated the differences between the conservative, practical policies of the UMWA and the more idealistic goals of industrial unionism espoused by the WFM.

Yet the Utah experience caused WFM idealism to wane. At the 1908 Annual Conference of the Western Federation of Miners, executive board member Roderick MacKenzie concluded that the union's attempt to take over the UMWA in the Utah and Colorado coal fields had failed; and since the UMWA had not abandoned the western coal fields, an unhealthy state of competition existed. MacKenzie predicted that it would take the UMWA "at least five to ten years to overcome the prejudice against their organization which was created in the minds of the miners in Colorado and Utah by the disastrous failure of the coal miners' strike of 1903 and 1904."[18] The situation was perilous. MacKenzie maintained, "Prejudice is so strong that even the lives of organizers of the U.M.W.A. are not safe among former members of that organization." Despite these handicaps, MacKenzie concluded that the Utah and Colorado miners could not benefit "without the full cooperation of the Western Federation of Miners with the United Mine Workers of America."[19] However, such cooperation was not forthcoming, and for the next decade the once thriving Utah union movement was engulfed by problems.

The demand for new miners to take the place of the strikers who had either left or not been rehired by the Utah coal companies and the expansion of the coal-mining industry in Carbon County coincided with a wave of Greek immigrants coming to America. Forced to leave Greece because of perpetual poverty aggravated by the collapse of the country's economy when the currant industry failed completely in 1907, Greeks looked with great hope to America—especially because earlier immigrants, clothed in their new American finery, had sent pictures home and had written of their great fortune. These personal testimonies of the promise of America were substantiated by advertisements in Greek newspapers. Throughout the Greek mainland and on Crete, papers carried notices for workers in America with guarantees of work and prosperity. The ads were placed by steamship agents who traveled to the remote mountain villages to secure clients for their ships leaving Greece and other European ports for America.[20]

The 1910 census registered the largest concentration of Greeks in America as living in the Mountain States with Utah's Greek population soaring from 3 in 1900 to 4,062 in 1910.[21] Utah became the destination for hundreds of Greek immigrants, primarily because of the influential Leonidas G. Skliris, the Greek Padrone or "Czar of the Greeks," as he became known throughout the Intermountain West.

According to one newspaper report, Skliris was the son of a former governor of the province of Sparta and a member of one of the most prominent families in Greece. For some undisclosed reason, he left Greece about 1895 as a stowaway, and after arriving in New York City, he sold flowers in the lobby of an office building in the financial district. Saving enough money to start west, he worked on a construction gang for the Illinois Central Railroad and then contracted to supply labor for the Cripple Creek Railroad in Colorado. Successful in this undertaking, Skliris became a labor agent for the D&RGW; and after moving to Salt Lake City, he became a labor agent for the Carbon County coal mines, the Utah Copper mine at Bingham, the Salt Lake Valley smelters, and other mining and railroad operations throughout the Intermountain West.[22]

Living in the luxury of the Hotel Utah, Skliris operated his business out of his headquarters and store located in the heart of Salt Lake City's Greektown at 507 West 200 South. After the first years of finding recruits in Greece, Skliris and his lieutenants were able to secure enough men from the Greek coffeehouses of Chicago's Halsted Street or from those in Denver and Pueblo, Colorado. Later, with a large number of Greeks under Skliris's control, the recruiting efforts outside Utah were no longer necessary.[23]

Under the padrone system, Skliris provided his countrymen with jobs as common laborers at a charge of between ten and twenty dollars. In addition, each laborer agreed that a smaller amount, usually a dollar, would be deducted from each month's pay and given to the padrone. In some cases, such as at Bingham, the men were required to trade at stores and businesses operated by Skliris. In the coal camps, the agents of the padrone were usually identified as "interpreters." Because of their ability with both the Greek and English languages and because of Skliris's agreement with the coal companies to supply men, the interpreters could control their jurisdiction with little interference. The padrone system worked especially well for the coal companies, where cheap labor could be provided, managed, and shielded from the radical ideas of unionism. Company officials frequently received kickbacks from the padrone from the money each immigrant paid to secure his job.

Although the Greek immigrants initially appreciated the padrone's role in securing work for them in America and in providing a means of communication with their English-speaking bosses, they quickly tired of Skliris and what they considered his heavy-handed methods of extortion. In 1911, Greeks at Bingham wrote an unsigned letter addressed to the "Utah Government, Salt Lake City, Utah." Unclear as to who had appointed Skliris labor agent for the Utah Copper Company, the writers complained that in applying for work with the Utah Copper Company, they were required to have "a letter of permit" from Skliris; when they sought such a letter, Skliris demanded fifteen to twenty dollars. Fearful that if they signed their names to the letter, Skliris would see that they were denied jobs wherever he could, the Greeks nevertheless volunteered to testify against Skliris if they could secure positions without his approval. Pleading for assistance in a passionate style of recently learned English, they implored:

> Do you think this is right for him [Skliris] to sell livelihoods to the poor workmen at exortion 20.00 dollars and to thus suck the blood of the poor laborers. Where are we? In the free country of Amerika or in a country dominated by a despotic form of government as under the suberenity [suzerainty] of the Sultant? For God sake release us from such assessment and let us be independent in this independent country and require [acquire] position through the foremen just like others.[24]

After receiving several complaints about Skliris's exploitation of the Greeks, Governor William Spry finally contacted the Utah Copper Company for an explanation. In 1912, R. C. Gemmell, the company's assistant manager, denied the charges against Skliris. Gemmell acknowledged that Skliris was paid a monthly salary as a labor agent for the Utah Copper Company but firmly stated that he was not permitted to collect any money from Gemmell's men. The Bingham situation was investigated by a Greek detective under Gemmell's employ, and he could find no proof that Skliris collected anything from the men or forced them to trade at a particular store. Moreover, the company, Gemmell claimed, notified all employees that they were not to pay money or give any presents to foremen, bosses, or anyone else. Acknowledging that Utah Copper had received signed petitions denouncing Skliris, Gemmell claimed after an investigation that those signing the petitions were not employed by the company but were butchers, bakers, barbers, and men who had worked for the company but who had been discharged for failure "in the performance of their duties." The complaints, Gemmell concluded, were instigated by Stampolas, a Greek storekeeper in Bingham and an enemy of Skliris.[25]

After the Utah Copper Company refused to acknowledge that abuses of the padrone system existed, the Greek miners turned to the WFM for help in getting rid of Skliris. On September 17, 1912, a strike was called at Bingham. Although the WFM was concerned primarily with higher wages and with recognition of the union, the first concern of the Greek strikers was to have Skliris fired as labor agent for the copper company.[26] Despite a strong denial that he was a padrone, Skliris resigned his position as labor agent for the Utah Copper Company a week after the strike began.

Discontent with Skliris and the padrone system had been manifest earlier in Carbon County. On June 16, 1908, George Demetrakopoulos, an agent of Skliris, was shot and killed in his room at a Winter Quarters boarding house. Demeter, as he was called by the non-Greek community, was "a single man about 30 years of age, well educated and posted on Greek history and modern sciences."[27] A reputed graduate of the University of Athens, Demeter, upon his arrival in the United States, edited a Greek newspaper in Chicago. He "later went to Salt Lake City with the avowed purpose of starting a Greek newspaper there. This plan miscarried, and he entered the employ of L. G. Skliris as interpreter."[28]

Demeter's assailant was identified as Steve Flemetis. Bad feelings had existed between the two for some time. Flemetis was described as a former employee of the Utah Fuel Company and it seems likely that the shooting occurred over Demeter's involvement in having Flemetis dismissed. On the day of the shooting, Flemetis obtained permission from the landlady to visit Demeter in his upstairs room. Demeter was working at his typewriter when Flemetis stepped into the room and shot him in the back of the head. Despite an almost immediate response by local authorities, Flemetis escaped. Other members of the Greek community were accused of conspiracy in the murder and of assisting in Flemetis's escape.

The death of Skliris's agent disrupted the functioning of the padrone system in the coal fields for a time. In July the Utah Fuel Company responded to attempts by interlopers to take over for the dead Demeter by posting notices warning Greek miners to beware giving money to interpreters for employment and instructing them to report all attempts to extort money.[29]

Nevertheless, Skliris succeeded in installing another agent, Thomas Panas, to replace Demeter. A "Black Hand" letter threatening Panas with death was sent to the new agent by a Greek miner at Winter Quarters who had been fired because of Panas.[30]

Although the "Black Hand" threat against Panas was not carried out, a disgruntled compatriot, Gregorios Pologeorgis, nearly succeeded in killing Skliris. Pologeorgis had operated a store in Scofield and, according to his account, had borrowed $100 from Skliris. The money was repaid to John Cardasis, a cousin of Skliris, who apparently worked at Scofield in the same capacity Demeter and Panas had earlier.[31] Pologeorgis was not given the paid note, and when he wrote to Skliris asking that it be returned, he received no reply. Consequently, he sought out Skliris in Salt Lake City. The two met in front of the Co-operative Furniture Company building at 39 South Main. An argument ensued, and when Skliris called Pologeorgis a foul name, Pologeorgis shot Skliris in the chest. Skliris subsequently acknowledged the note had been paid, but by Pologeorgis's partner because, according to Skliris, Pologeorgis had refused to pay the debt. Pologeorgis pled guilty to a charge of assault with intent to commit murder and was sentenced to an indeterminate term in the state prison.[32]

The Greek miners' frustration with the padrone system matched their bitter feelings against the coal companies. In Kenilworth, Greeks charged the Independent Coal and Coke Company with cheating at the weighing scales and with favoring American miners. On Thursday, February 2, 1911, the morning shift of Greek miners refused to enter the mine until the weighing scales were adjusted to reflect the accurate amount of coal mined. They also demanded to select a checkweighman. Superintendent Thomas Bell met the men and agreed to balance the scales, to record the weight of each loaded car, and to permit them to select a checkweighman. According to the coal company officials, the Greeks agreed to these concessions, however, "some of the ringleaders . . . acting in an insulting and boisterous manner" refused to go to work that day. Consequently, fifteen Greeks were fired and escorted out of the camp. They secretly reentered Kenilworth and, meeting with Greek, Austrian, and Italian miners, drafted a letter to the Independent Coal and Coke Company in which they again called for fair weights and for their own checkweighman. The miners also demanded that the discharged men be reinstated and that miners be allowed to trade wherever they wanted. The miners vowed they were prepared to strike "until we get satisfaction."[33]

The vexed miners did not receive an acceptable answer and concluded they would prevent the mines from working until their grievances were resolved. Accordingly, late Sunday evening, February 5, they took up positions among the rocks surrounding the camp. Two hours before daybreak, nearly all the Greek miners had left camp, and fires were observed in the hills north and south of Kenilworth. Early Monday morning, after the American miners finished their breakfast and began walking toward the mine entrance, strikers, concealed among the rocks above the town, announced their resolve with a volley of pistol and rifle fire aimed at the miners and nearby buildings. Two bullets passed through the residence of Superintendent Thomas Bell, just missing his wife. Numerous shots hit the mine company's main office, and as the firing continued "practically every house where Americans were known to live was penetrated by pistol and rifle bullets."[34] Women and children found cover in cellars and behind rocks while miners and company officials located rifles and ammunition and formulated plans to break the siege. Under the leadership of Thomas Jackson, a guard for the Independent Coal Company, a group of miners moved out of the town and opened fire on the strikers, forcing the Greeks to flee for the safety of the high rim rocks. However, as the miners pursued the strikers up the steep mountain, one Greek took careful aim with his rifle and shot Thomas Jackson in the chest, killing him instantly.

Thomas Jackson was the first of several men to die in the Carbon County labor disputes. Jackson's death was soon followed by a striker's—Steve Kolasakis. When the body of Thomas Jackson was carried into the camp hospital, Robert Jackson seized his dead brother's rifle. Accompanied by Henry Williams, his half-brother, Robert Jackson started in pursuit to avenge the death. In full view of those in the camp, who feared that the two men would soon share their brother's fate, Jackson and Williams made their way up the mountain. As the shots continued from the summit, Robert Jackson sighted in on one Greek and fired. "The Greek rose, seemed to grasp at his forehead and then fell back dead, simultaneously dropping his revolver which clattered down the mountainside."[35] The dead Greek was identified later as Steve Kolasakis.

The battle ended with fatalities on both sides. Late Monday afternoon, February 6, a contingent of forty men sent by train from Salt Lake City to Helper met a local posse of twenty men led by Sheriff Thomas Kelter and together made their way to Kenilworth. Kelter concluded that the Kenilworth Greeks had been assisted by their countrymen from other camps, especially Castle Gate. The sheriff announced that he would defend Kenilworth from further attack and at the same time search for the Greek leaders, probably making their way to Helper. It was rumored that many Helper businessmen were giving at least tacit support to the strikers because the Kenilworth miners were forced to patronize the company store rather than trade with independent merchants.[36] Kelter's resolve was firm as he pledged "to run down all the Greeks responsible for the outrage if it takes weeks."[37]

The search for the strike leaders did not take long. Monday afternoon, February 13, Henry Kallos and John Matsos, the alleged ringleaders, were arrested, manacled, placed under heavy guard, and arraigned before Kenilworth Justice of the Peace E. J. Lambson on charges of intent to commit murder by assault with a deadly weapon. The arraignment was held in the Kenilworth Mercantile where cracker boxes, rolls of rope, and kegs of nails served as courtroom seats. Interpreters for the two Greeks proclaimed that the men were not guilty. When it was announced that the preliminary hearing would be held later, emotions flared and "angry eyes were turned towards them [the Greeks]." Violence was prevented, in the nativistic assessment of a *Deseret News* reporter, only because of "the superior intelligence of the American, his respect for law and order prevented even there the partial avengance of the crime."[38]

Three other Greeks—John Poulenakis, George Monakis, and Nick Karamousis—were arrested, and warrants were issued for the arrest of Andrew Stellos and John Makis, who were allegedly wounded during the shooting. Karamousis was held only one night and was released Tuesday morning when he was able to prove he had not participated in the attack. Later Monakis and Poulenakis were discharged. Stellos and Makis, according to Sheriff Kelter, were taken by sympathizers to Helper and were hidden. The two men were not apprehended despite a thorough search of the Greek coffeehouses in Helper.

The death of Thomas Jackson and the violent manner in which the Greek strikers sought to deal with the situation brought public demands that something be done about the Greek "menace." In his report to Governor William Spry, State Mine Inspector J. E. Pettit noted that "it has been customary in our different camps to keep an equal per cent of foreigners, but at the present time there seems to be a perponderance of the Greek element."[39]

There was good reason for the perceived need to maintain an equal percentage of foreign miners. Coal company officials recognized the practice as one of the most effective measures for discouraging the growth of union sentiment. One Kenilworth official disclosed that the Independent Coal Company "had sought to guard against union troubles by installing workmen of different nationalities, Greek, Italian, and American, as there was less chance of combinations among the men than when one nationality was exclusively employed."[40]

It was taken for granted that those who had participated in the strike and the shooting would not be taken back by the Independent Coal and Coke Company. Despite his acknowledgement that "there are many peaceable Greeks who are efficient workmen," Superintendent Bell openly speculated whether he would permit the sixty Greek miners who had nothing to do with the strike, to return to work.[41] A *Salt Lake Tribune* reporter echoed the strong and familiar sentiment "that the time is ripe for the inauguration of a system that will result in the employment of a much larger percentage of Americans in the mines than are used at present."[42]

Prominent Greek leaders, in an effort to restore better relations between the coal companies and their Greek countrymen, traveled from Salt Lake City to Kenilworth. Although no documentation is available, presumably one of the Greeks was Leonidas Skliris or one of his subordinates. The threat not to employ Greek miners because of the Kenilworth trouble posed serious problems for the Greek padrone, who had been instrumental in supplying men for the mine since it opened in 1906. The Carbon County situation challenged Skliris's power not only there but in the Intermountain West as well because growing anti-Greek sentiment threatened to discourage the employment of Greeks outside Carbon County.

Charles Soter, another prominent Greek from Salt Lake City, was given the task of soothing the public outcry against the Greek miners. Newspaper reports noted that Soter was sent to Kenilworth to act in a triple capacity: special deputy, interpreter, and mediator-peacemaker between the coal company and miners. Soter was especially anxious to discourage Greeks in other coal camps from joining the Kenilworth strikers. His tactic for restoring relations between the Greek miners and both the coal companies and the public reflected the extent to which old-world animosities and antagonisms had followed Greek immigrants to America. Soter placed sole blame for the trouble on the Cretans announcing, "They are easily led into trouble and will follow agitators without question." The mainland Greeks, in contrast, he promised, would do what they were told and avoid any trouble. Pushing for reinstatement of the Greek miners, Soter declared if they were "allowed to go back to work, it is well; if they are discharged, they will go away and create no disturbance."[43]

Soter succeeded in his mission. Nearly all the Greek miners were rehired, even many who had participated in the shooting. Several people questioned the lenient hiring practice, including J. P. Curtain, surface superintendent, who resigned his position declaring there were enough natural dangers in coal mining without the threat "of being slaughtered by Greek miners. I know that they are awaiting an opportunity to kill mine officials and when I resigned I figured myself a likely candidate."[44]

Curtain's fears were not without justification. A month later, the *Salt Lake Herald* and *Sun-Advocate* noted that "anonymous 'Black Hand' letters ordering them to leave the community under penalty of death have been received by Superintendent Thomas Bell and Chief Clerk Hogan . . . at Kenilworth." Bell and Hogan defiantly promised to "stay on the grounds and die with their boots on if necessary."[45]

The supreme sacrifice was not required of the two Independent Coal and Coke Company officials. However, the short-lived strike had left two dead men. Thomas Jackson and Steve Kolasakis were the first recorded deaths directly attributable to a labor-management confrontation in the Utah coal fields.

No mention was made of union organization as part of the 1911 strike. The strikers sought to redress their grievances as they would in the "old country." Vendettas required only individual action; unions were not necessary. The strike at Kenilworth lasted only four days; and if the UMWA had been interested in sending organizers, it would have had too little time to take advantage of the situation.

Between 1911 and 1913, efforts to redress the workers' grievances were sought first through local political initiatives rather than through union affiliation. During this period, Carbon County found itself in the mainstream of the nationwide reform movement. Characteristic of the national movement, Carbon County reformers felt that change could come best through the political process. Accordingly, the 1912 election in

Carbon County saw, in addition to the traditional Republican and Democratic candidates, nearly a full slate of local candidates representing the Socialist and Progressive parties. The Socialist party was greatly strengthened by the staunch support of the *Carbon County News*, a Price newspaper started in April 1907 and acquired by W. C. Benfer in August 1911.

Some time before the Kenilworth strike, Felice Gigliotti, a stonemason and professional organizer for the Socialist party, succeeded in establishing a local organization in Helper. Meetings were held every second and fourth Sunday at Lowenstein Hall in Helper. By the late summer of 1912, the organization had grown sufficiently to field a nearly complete ticket of local candidates for the November election. The 1912 campaign was launched with a Labor Day parade in Helper. The procession, which extended the full length of Helper's main street, included a large banner bearing the pictures of Eugene V. Debs and Emil Seidel, the Socialist party candidates for president and vice president. Other banners interspersed among the marchers "indicated that labor would vote the socialist ticket."[46] After the parade marched along the Helper streets, the participants gathered in front of the post office to hear a selection by the Helper band and a political speech by W. A. Jameson, Socialist candidate for county sheriff. In his speech, Jameson hammered home the importance of workers being organized to protect their interests. Because the interests of the workers were opposite those of the corporations and because all political parties except the Socialist supported the interests of the corporations, it would be a mistake for union supporters to "sustain . . . men in office who have prohibited labor from organizing." Jameson concluded his speech by admonishing the "workingmen to read, think and study and support the party of their class."[47]

From September 5 until the November election, the *Carbon County News* devoted space to political advertising. In reality the section contained only the Socialist party views because no other party used the space. In its first column, the Socialist party bluntly lashed out against the tyranny of the capitalist system; it promised workers protection in future labor disputes if the Socialists were elected. The article declared:

> By owning the jobs upon which the life of the worker depends, the corporations are able to dictate who all government officials shall be; thus are they able to control the administration of affairs in their own interests and against the workers. They have no scruples about breaking the law nor resorting to any brutality necessary to subjugate the working class.

> The present oppressive conditions make it necessary for the working class to cancel all relations with the three capitalistic parties and become identified with their own party the socialist party.
>
> This socialist party has arisen as a protest against the confiscation and dividing up of the earnings of labor with the bosses. It stands for a fair deal to all, privileges to none and against the judicial rape perpetrated by republicans and democrats alike in their ignoring of the constitutional rights of labor.
>
> The administration of the affairs of Carbon county is no exception to the rule. Wherever you find capitalism you find brutality and graft.
>
> The affairs of our county have been greatly mismanaged by members of both the republican and democratic parties, by dishonesty and incompetency. . . .
>
> Our candidates, if elected, will see that you are not beaten up by thugs when on strike, and that you are not deprived of your liberties without due process of law. We expect to enforce all labor laws and to disarm all troublesome and law-breaking deputies and gun men.[48]

The Socialists challenged the Progressive party's labor policy. When the Progressive party announced its slate of officers and outlined its platform, the *Carbon County News* could not find "one word approving the fundamental right of labor to organizing, or even recognizing this right." The paper proclaimed that although the Progressive party promised social and industrial justice it did not promise industrial democracy. Therefore, the only thing the working class could expect from the Progressive party promise of industrial justice was "that which the benevolent corporation is prone to administer thru its welfare department."[49]

Nevertheless, coal company officials felt a legitimate threat from the Progressives, and in Sunnyside the party was denied the use of a public hall where a rally was scheduled. When the Progressive candidates suggested a street meeting, Deputy Sheriff Hyrum Wilcox informed them that if they made any statements critical of the Utah Fuel Company, "they would be fired out of town."[50]

The election results indicated a relatively strong showing by both the Progressive and Socialist parties. The Socialist vote averaged about 17.3 percent of the total vote; the Progressives captured 23.4 percent of the vote. The Progressive candidates for Four-Year County Commissioner, Joseph R. Sharp, and Two-Year County Commissioner, W. T. Hamilton, won their contests.

In the reform tradition of the Progressive party, the two commissioners sought to wrest political control of the county from the coal operators.

They were aided in their efforts by the *Carbon County News* and *The Progressive*, a weekly newspaper published in Salt Lake City. *The Progressive* was especially vocal in attacking the Utah Fuel Company's control of county government and of the lives of the company's employees.[51] In the months following the election, the county commission battled over dismissing the deputy county clerk, appointing a "special Council" when the county attorney purposely failed to attend commission meetings, refusing the *Eastern Utah Advocate*—mouthpiece for the coal company interests—the county's business for printing services, and rejecting Sheriff Thomas Kelter's request that Hyrum Wilcox be appointed a special deputy to serve without pay (Wilcox was a long-time employee of the Utah Fuel Company and, as noted, had refused the Progressive candidates permission to meet at Sunnyside prior to the election).

Despite the anticapitalist, prolabor rhetoric of the *Carbon County News*, the Utah Fuel Company carried a four-dollar-a-month ad in the paper until January 30, 1913. The apparent reason for cancelling the ad was the stand taken against the Magnolia Trading Company, described by the *News* as the "saloon owning wing of the [Utah] Fuel Company."[52] The *News* held that until the election of the Progressive commissioners, the Utah Fuel Company (through the Magnolia Trading Company and its control of the county officers) had been able to maintain a monopoly of the liquor business in the county. The new county commission granted liquor licenses to those challenging the company's domination of liquor and mercantile traffic near some of the coal camps. In Sunnyside, the license was given to Fred Paternoster and J. Scarpino, operators of the newly constructed Golden Rule Store, which by its very name offered miners relief from the tyranny of the company store. Mark Braffet of the Magnolia Trading Company sought to keep residents from patronizing the Golden Rule Store by fencing the Utah Fuel Company property. When the county commissioners learned of the fencing project, they quickly held a meeting and approved a county road from Mounds to Sunnyside. This required the coal company to leave an opening in the fence for the proposed road.[53]

In retrospect, the efforts of the Socialists and Progressives served more to harass the Carbon County coal operators than effect any long-lasting change. Within a few months, the rhetoric against the coal companies subsided, and the Utah Fuel Company once again took out its four-dollar-a-month ad in the *Carbon County News*. The political threat no longer seemed ominous, and the coal companies turned to meet another challenge from the left: this time from the Industrial Workers of the World.

One of the most colorful union organizations to develop in the United States was the Industrial Workers of the World. Organized in Chicago in 1905, the IWW, or Wobblies as they were often called, represented socialists, anarchists, radical miners, and revolutionary industrial unionists. Its purpose was to overthrow the capitalist system and replace it with a "cooperative commonwealth" of workers.[54] The zenith of IWW activity came in 1912 when strikers at the Lawrence, Massachusetts, textile mills won a 7.5 percent wage increase.

The Lawrence victory encouraged the IWW in other regions during 1913. The West proved a stronghold for the IWW, especially among the railroad construction workers and in the lumber and mining camps. In Utah, IWW organizers arrived at the camps of the Utah Construction Company, located between Tucker and Soldier Summit, early in June 1913. By June 11 the IWW leaders had persuaded many of the 1,500 to 2,000 workers employed by the Utah Construction Company to strike. The men were laying a double track across Soldier Summit for the D&RGW. The five demands of the strikers included: a pay increase of twenty-five cents a day; a nine-hour workday instead of ten; elimination of the dollar-a-month hospital fee which was, at times, required more than once a month if the workers moved from camp to camp; abolishment of the company practice of paying men with checks and then assigning a company agent who would cash the checks at a discount; and improvement in sanitary conditions including free baths, bunks, and bedding.[55]

D. H. Jansen, the IWW spokesman, presented the written demands to the company. A. H. Christensen, who was in charge of work for the company, refused to look at the statement and returned it to Jansen declaring that those making the demands did not work for the company but were transients who had "never worked here and have simply come here for the purpose of causing trouble."[56]

W. H. Wattis, president of the Utah Construction Company, was optimistic that the strike would be settled quickly because of the satisfaction that had existed previously in the construction camps. The tranquility had been disturbed, he noted, with "the invasion of the agitators." Wattis described the Industrial Workers of the World as "tramps" who had stopped off in the construction camps before the company could protect itself "and [who] by their speeches stirred up a number of laborers who love excitement better than work. These men in turn either convinced others that they were being mistreated or forced them to leave."[57]

Wattis denied that the company had a representative to cash checks at a discount, but he maintained that since the men were paid with ordinary bank checks, many of them, in a hurry to get spending money, did cash their checks at a discount. The president asserted that men were required to pay the one-dollar hospital fee each month, no matter what the circumstances, and that if they left before the end of the month, they were reimbursed a portion of the fee. Furthermore, Wattis remained adamant in refusing to reduce the workday from ten to nine hours.

Shortly after the strike call, many workers left the camps. Although reluctant to man the picket lines in support of the strike, they did not want to be branded as scabs by those on strike. One man explained: "I have been a laboring man in all camps around the country and I have never been called a scab yet, and I am not going to be. I'm going to get out."[58]

Threats were made against those who wanted to work; men were driven from their working places by strikers armed with pick handles. In traditional Wobbly rhetoric, one leader declared, "The IWW would win the strike if they had to cut the harnesses off the horses of the Utah Construction Company, dump their scrapers in the fills and tie dynamite to the horses tails."[59] Official statements were much more restrained, and IWW leaders urged their men to avoid violence.

Because of the nature of the construction work, a large supply of dynamite was on hand, and it was feared that the strikers would seize the explosives and perhaps blow up sections of the work completed or destroy the expensive steam shovels essential to the work. At one of the camps, a pump was disabled. In Price, transient IWW members were blamed when a mail sack was stolen and ripped open.

The strikers knew that deputies would be sent to deal with the situation. Shortly after the strike began, a notice was posted on the IWW bulletin board announcing that four sections of cavalry from Fort Douglas were en route to Tucker and that twenty strikers had been killed by "whiskey crazed deputies."[60] Strike leaders denounced the notices as false rumors, which they proved to be. Utah Construction Company officials were anxious to remove the strikers from the construction area for two reasons. First, they feared that sabotage might be carried out against company property. Second, with the agitators and sympathizers gone, it would be much easier to bring in new men willing to work under the conditions against which the strike had been called. On June 11, the day after the strike was called, deputy sheriffs rounded up 128 men at Tucker, loaded them into two railroad passenger coaches, and started for Provo. As the train approached other construction camps, it stopped, and the deputy sheriffs invited other strikers to board the train. When the strikers refused, the deputies fired warning shots above their heads.

Five strikers, J. F. Morgan, E. Raywon, L. Weaver, A. C. Ridgeway, and A. J. Rice were taken from the special train, arrested, and placed in the Provo jail on charges of disturbing the peace, inciting labor trouble, and of destroying property belonging to the Utah Construction Company.[61] While in jail, the men made their presence known by shouting and singing until they became hoarse. Later in the evening, four other IWW members—George McCue, Frank Duffy, Frank Eubanks, and Frank Rice—joined their five comrades in the Provo jail.[62] The other IWW passengers aboard the special train were allowed to go free. Some made their way back to the strike zone while others moved on looking for work elsewhere.

The IWW maintained that it had the territory between Denver and the Pacific Coast under control and that the Utah Construction Company would be unable to bring in strikebreakers. Utah Construction Company officials showed no visible concern with the IWW tactics. When questioned as to how they would distinguish between those honestly seeking employment and those IWW sympathizers sent to the strike area to further disrupt work, Christensen explained it would be simple—all they would have to do is look at the applicants' hands. "There are no full fledged Industrial Workers of the World with work callouses on their hands. All we have to do is select men . . . who bear the evidence . . . in their toil-worn hands."[63]

IWW sympathizers did arrive from Colorado; however, they were identified and were refused employment. The IWW assertion of having the entire area under its control did not prove true. Men were found to take the places of those who remained on strike, and many who left when the strike began returned after special deputies were sent by the company to protect those who wanted to work. Those arrested were tried in Provo and were given jail sentences ranging from sixty to seventy-five days. When A. C. Ridgeway was brought before the judge, he confessed that he had incited the strike; but in a dramatic statement of the IWW hope for change Ridgeway declared, "Some day there would be enough of his way of thinking in the country so that he might be sitting in the prosecuting attorney's chair."[64]

After the sixty days in jail, the IWW leaders traveled north to Salt Lake City and immediately became involved in a riot. When James F. Morgan mounted the speaker's box to rail against the Utah Construction Company and other Utah capitalists for mistreating workers, he was knocked from the stand and was beaten over the head with a revolver by Axel Steele, who was in charge of company guards for the Utah Copper Company. In the melee that followed, shots were fired, the IWW banner was replaced by the American flag, and the fire department was called to quell the rioters with water hoses. Eight men were arrested. F. J. Clifford, a former guard at Tucker, was shot in the leg.[65]

To the coal operators in eastern Utah, the IWW strike at Tucker and the riot in Salt Lake City portended IWW agitation in their coal fields. Many of the men employed on the railroad construction project probably had experience in the coal mines. The summer months were usually extremely slow in the mines; because of the nearness of the construction project, the lack of work in the mines, and the demand for baths (which coal miners were provided with daily after leaving the mines), it is likely that a portion of the strikers were also Utah coal miners. Rumors circulated that the IWW would attempt to organize the coal mines of Utah and Colorado.[66] The IWW requested permission to hold street meetings in Helper and Price. When permission was refused, the IWW announced it would hold the meetings anyway. However, the meetings were never held and the fear faded of an IWW organization in the Utah coal fields.

IWW efforts would continue with little success in Utah's metal mines and smelters, yet Utah and the Industrial Workers of the World would share center stage in the wide-playing epic of labor conflict. As the IWW retreated from the mountains of the Wasatch Plateau in the summer of 1913, the thirty-four-year-old IWW organizer and songwriter Joe Hill boarded a train in Los Angeles headed for Salt Lake City. Stopping to earn a "grubstake" before continuing to Chicago, Hill, a native of Sweden, found work in a Park City mine owned by the Silver King Coalition Mining Company. Taken ill, and then dismissed by the mining company, Joe Hill was arrested for the murder of a Salt Lake City grocer, John G. Morrison, on January 10, 1914. Stoutly maintaining his innocence but doing nothing to counteract a barrage of circumstantial evidence, Hill was convicted and sentenced to death. Executed on November 19, 1915, by a firing squad at the Sugarhouse prison in Salt Lake City, Joe Hill became a martyr for the IWW. Carried out in the face of vocal national and international protest, the execution capped Utah's reputation as the nation's hotbed of antilabor sentiment.[67]

While the Utah coal operators were concerned with the threat of IWW agitation among the miners, southern Colorado coal operators were attempting to thwart a determined organizational effort by the United Mine Workers of America. The UMWA had given the Colorado effort top priority, and in December 1912, Frank J. Hayes, vice president of the coal miners' union, arrived in Colorado to assist the organizing efforts.

By late summer 1913, the union was sufficiently strong to issue a strike call for September 23. The demands included higher wages, better conditions, and union recognition.[68] By the end of September, it was estimated that between 90 and 95 percent of Colorado's 14,000 coal miners were on strike. Strikers evacuated the company camps and gathered in tent colonies. A month after the strike began, the state militia was called out.

The strike was characterized by arrests and violence. In January Mother Jones, by then eighty-four years old, was arrested and placed under guard in the Trinidad hospital. When a group of women organized a demonstration parade in support of Mother Jones, they were dispersed by a mounted saber charge. Two months after this event, in mid–March 1914, another mounted charge at Ludlow, Colorado, resulted in the death of two strikers.

The most tragic event of the strike, and indeed of all labor relations in the West, was the death of two women and eleven children at Ludlow on April 20, 1914. The Ludlow Massacre became the nationwide symbol of coal company brutality, and in 1915 the UMWA erected an impressive monument in memory of those killed at Ludlow. The Colorado strike lasted more than fourteen months but when it was finally called off on December 10, 1914, the miners had failed to obtain the important goal of union recognition. In addition, most of their places in the Colorado mines had been taken by strikebreakers.

Although Utah was a part of Colorado's UMWA District 15, it was not included in the September 1913 strike call. Union organizers had focused exclusively on the southern Colorado coal fields and did not push into the Utah area until just before the strike was called. Just after the strike began in Colorado, the Utah state mine inspector, J. E. Pettit made a hurried inspection of the Utah situation and found conditions normal; Utah miners were reportedly satisfied with conditions and were not taking much interest in the Colorado strike. Pettit went on to report that "a number of agitators representing the I.W.W.'s Western Federation and United

Mine Workers, have made attempts during the past two weeks to get into the Carbon-Emery County camps." The individuals, he concluded in a letter to Governor William Spry, had made their headquarters in Helper and Price but were "being closely watched by the mining and county authorities."[69]

In Helper, 100 pounds of giant powder and fifty sticks of dynamite were found hidden in the D&RGW coal supply. Sympathizers with the coal strikers were suspected of intended sabotage.[70] The *Eastern Utah Advocate* noted that the mine superintendents of the Consolidated Fuel Company and the Utah Fuel Company had received written threats that seemed to be connected with the Colorado strike.[71]

When the strike began, Colorado coal operators sought to recruit workers from Utah. One man, who requested a monthly salary of $175 to leave his position at Castle Gate, reconsidered and turned down the Colorado offer because of the labor trouble.[72] As the strike dragged on, men who had participated in the Colorado strike were reported to be arriving in the Utah coal fields in large numbers.[73] It was promised that the UMWA would attempt to organize the Utah coal fields in September 1914. Nevertheless, the union had its hands full with the Colorado strike and the anticipated organizational campaign in Utah did not occur. Later, miners who came to Utah from the southern Colorado coal strike would recall and equate the experiences of 1913 and 1914 with those of the Utah coal fields.

In April and May 1915, coal miners at the Neslen mine of the American Fuel Company of Utah went out on strike in an effort to collect back pay. Neslen, located east of Green River at the foot of the Book Cliffs in Grand County, was one of the most isolated coal camps in the state. The coal company lacked the financial stability of the larger companies located in the Carbon County area, and the two walkouts demanding back pay for five months were unique in the history of the eastern Utah coal fields. In a letter written by William Bell, mine superintendent at Neslen, to his friend State Coal Mine Inspector J. E. Pettit, Bell outlined the situation in the Neslen camp and indicated more than just a problem in meeting the payroll.

> The work was very slack here from Christmas to about the middle of Feby. I closed down No. 3 mine and layed off about 50 men who called for their time. A large percent was Itallians and the night they was to leave town they got a lot of Grand county Probition whiskey and got fighting and murdered one man wich caused us considerable trouble.

> I have got all those out of camp that I thought to be rather bad actors and things are quitened down again. . . .
>
> I have been since November and never had a rail sent me four months and was a long time and had no Giant Powder, I got along fairly good with the Black Powder. Now the Black Powder is all gone and some Giant Powder came a few days ago of two or three different grades some sticks of 8 in and some 1½ in and no Black powder. Miners are dissatisfied on all sides and kick on the high price.
>
> In regards to things being better I do not see much change. We had no butter for dinner at the club on Sunday and not seen any fresh meat for a week only once or twice we had fresh pork. Mostly ham and eggs or bacon and eggs that once was fresh. People with families cannot get shoes and clothing or any underwear as there is none in stock. In regards to pay day they have not paid yet for December, Jany, and Feby, and the people never knows when to expect a pay. There will be a few hundred dollars sent down enough to pay about ⅓ of the men and that makes it worse and causes a feeling of dissatisfaction among those not paid. People with families has a hard time of it. Cannot get what they want in the store neither to eat nor to wear and no money to send any other place. I have a hard time to keep the men to work under these conditions and am very likely to have them all walk out any day if the pays are not more prompt and better service at the store.[74]

Six weeks later, the situation had not improved and the miners refused to work until they were paid. Bell once again explained the dire circumstances at Neslen and clearly sympathized with the coal miners.

> The company sent down $1000.00 one thousand dollars today wich is only about enough to pay one third of the men for Jany pay and that will only cause more discord with the men; quite a few of the miners have called for their time already and are making a kick because they cannot get them cashed at the Bank. If a man does not do his work and he gets fired then he cannot get his pay, and with the company we have had a few of these cases. The men with families cannot get half enough to eat and wear because there is nothing in the store and they cannot get their pay so as to buy goods any other place. The company is now owing all employees for Jany, Feby, March, and here is the 10th of April. The Doctor quit on the first of April because he could not get his salary and not any drugs or appliances to work with, not even a bandage gauze or cotton if an accident should occur.[75]

William Bell convinced the men to return to work. However, when he left for a visit to company headquarters in Salt Lake City, the men went out on strike again.[76]

By the fall of 1915, the company was still in financial difficulty. Bell complained about the cheap oil sent by the company for the coal cutters, which caused the machines to clog and burn. He also noted that when rush orders were placed for parts they were sent by freight instead of express causing coal production to be cut by several hundred tons. Bell expressed concern with company officials' requests that more Greeks be hired because they would work for less money. Bell argued, "They [the Greeks] will not set timber and keep their room safe and I don't want that class of miners."[77] The company remained two months behind in paying its miners, and in October 1915, it instituted a pay reduction ranging from 12 to 20 percent.

The financial instability of the coal company burdened the miners in numerous ways. One miner, Tony Virturi, who was injured in the mine, found he was denied hospital care at Grand Junction because the company had not paid its last bill. Virturi had not been paid in months; and after being denied care at the hospital, despite the fact that the company had consistently deducted hospital fees from his pay, the injured miner was forced to borrow money to return to Neslen.[78]

The response of the coal miners at Neslen to the financial difficulties of the American Fuel Company, although marked by complaints and short-term walkouts, indicated a remarkable degree of tolerance and long-suffering, a response that would not be characteristic of a later day.

Chapter VI

A New Opportunity: World War I and Afterwards

A new and dynamic era of growth began in 1916 bringing with it the opportunity for important shifts in the labor relations of the coal fields. The demands for coal caused by the Great War in Europe were enormous. Coal was needed to fuel warships and transport ships and to run the railroads; it was essential for manufacturing the steel used in armaments. In addition to the wartime need for coal, a severe winter in 1916–17 led to increased demands for fuel by consumers. By May 1, 1917, the price of coal had increased from $2.40 to $3.50 a ton, and it was rumored that by winter the price would reach $8.00 a ton.[1]

Utah coal producers faced two problems in meeting the demand. First, a shortage of railroad cars to haul the mined coal plagued coal operators throughout the war. Second, the need to secure enough men to mine the coal also troubled the coal companies. The United States Fuel Company sent a representative into the northern Colorado coal fields to secure men for its Hiawatha mine in Carbon County. Although the company hired several, it noted that the men were "the most radical union type in that district."[2] In Sunnyside the Utah Fuel Company erected a Japanese Amusement Hall in an effort to secure and maintain the services of Japanese miners.[3]

Substantial pay increases were granted, and by May 1917, the daily wage scale averaged four dollars. In most instances, the wage increases were granted before the miners resorted to strikes; however, difficulties arose on several occasions when workers demanded more pay. In the last week of January 1917, Kenilworth miners won wage increases from the Independent Coal and Coke Company when they threatened to strike.

Two weeks later, 200 miners at Castle Gate struck the Utah Fuel Company.

During the strike, David Morrison, an employee of the Utah Fuel Company, was killed when a runaway train of loaded coal cars crashed into the electric motor he was driving. It was rumored that the loaded mine cars had been turned loose by Greek and Italian strikers.[4] The Utah Fuel Company hushed the news of Morrison's death in an effort to keep the strike from getting out of hand.

Unlike other strikes when company officials refused to talk with the strikers or even acknowledge receipt of written demands, this strike was different. J. S. Thompson, general superintendent of the Utah Fuel Company, met with the strikers, treated them with the utmost consideration, and granted them "about everything they asked for."[5] The increases amounted to five cents a ton for digging coal, which raised miners' pay from thirty-seven to forty-two cents per ton or about forty cents a day for those paid a daily wage. To avoid similar walkouts in the other Utah Fuel mines at Winter Quarters, Clear Creek, and Sunnyside, the company quickly extended the Castle Gate agreement to cover these mines. Following the Utah Fuel Company, the Cameron Coal Company, Spring Canyon Coal Company, and Carbon Fuel Company granted similar increases.[6]

But beneath the surface, there were indications that management's point of view remained unchanged. A week after the Castle Gate settlement, miners at Standardville in Spring Canyon struck the Standard Coal Company for a wage increase. F. A. Sweet, president of the company, declared that the Standardville miners had been given a 10 percent wage increase three months earlier and that they signed an agreement stating they would not ask for additional raises within a year. Sweet also noted that the recent Castle Gate agreement only raised wages at the other mines to equal what the Standard Coal Company had been paying. The strike, according to Sweet, "was apparently the first step in a movement to make the companies concede another blanket increase."[7] On this occasion, Sweet held his ground. The strikers were dismissed but allowed to return to work when they agreed to mine coal at the old rate.[8]

In June 1917, another effort was made to win a wage increase when 200 miners struck the Jesse Knight mine at Storrs in Spring Canyon. In addition to a wage increase, the miners demanded that the company implement the promised semimonthly pay schedule and discontinue the practice of paying the miners at the store owned by George A. Storrs. This practice had all but forced the miners to trade at the company store. The strike lasted four days. J. William Knight, general manager of the company, met with the strikers, agreed to implement the semimonthly pay schedule, change the pay station from the company store to the company office, and investigate the pay schedules of the other Utah coal companies to make sure that the miners were paid the same as miners in the other camps.[9]

Three months later, in September 1917, J. William Knight was again called to Storrs to handle a strike. This time the motormen were on strike protesting the treatment by the mine superintendent, A. E. Gibson. Once again, the strike closed the mine for four days. Knight, however, was not as generous as before and refused to meet the motormen's demand that Gibson be fired. Instead, they were discharged.[10] Strikes for increased wages were avoided at this time when the Utah Fuel Company raised wages an average of 10 percent.[11]

In retrospect, 1917 was a good year for Utah coal miners. Wages increased as the coal companies, anxious to meet production schedules, acquiesced to demands in a manner unique in the history of the Utah coal fields. Circumstances for the establishment of a miners' union were ideal in 1917; however, union recognition was never mentioned in the brief labor disputes during the year. On February 8, 1917, the national officers of the UMWA announced that an active organizing campaign would begin in Colorado, Utah, and New Mexico.[12] However, the campaign apparently did not begin in 1917; at least no mention of organizers in the Utah coal fields during the year has been found. Perhaps, too, the manner in which the coal companies quickly met the coal miners' demands negated, in the miners' eyes, the need for a union to wring concessions from the usually reluctant coal companies. Nevertheless, although the UMWA missed a propitious opportunity in 1917, by the fall of 1918 it was active in establishing local unions in the Utah fields.

Organizers for the UMWA reentered the Utah coal fields in July 1918. John McLennan, international representative for the union, established an office in Helper and hired several local organizers to work in the various camps.[13] Shortly after the Helper office was opened, McLennan left for a trip to Denver. The office was left in charge of Christ Bozas, a Greek organizer for the UMWA. During McLennan's absence, the Helper marshal ordered Bozas to close the union office and leave town. Bozas refused and was arrested by the Carbon County sheriff a few days later on charges of vagrancy. McLennan quickly intervened but not before Bozas spent nineteen hours in jail.[14]

Nevertheless, to many, the union appeared on the verge of victory. It was reported that national president, John L. Lewis, had taken a personal interest in Utah as reflected in his choice of John McLennan, former District 15 president and leader of the 1913–14 Colorado strike, to conduct the campaign in Utah. McLennan's tactics were successful. When mine officials at Latuda denied him permission to hold a meeting on company property, he drew the men to a snow-covered hillside a half-mile outside the company town. As the snow fell, McLennan held a night meeting illuminated by a log fire. At the end of the meeting, every man joined the union. The *Utah Labor News* enthusiastically, but prematurely, announced:

> The United Mine Workers organization is now firmly established in Utah. After thirty years of struggle, sacrifice and effort the coal miners of this state are about to celebrate their emancipation and enter into the enjoyment of all the rights and privileges that have been secured by the coal miners of other states through the instrumentality of their union.[15]

The coal companies strongly opposed union efforts to organize. On October 11, 1918, A.E. Harvey, secretary of the Utah Federation of Labor, noted in a telegram to William B. Wilson, United States secretary of labor, that the Carbon County fuel companies were importing "Colorado gunmen." Harvey requested that federal officials be sent before violence erupted.[16]

In Sego, Utah, thirty-seven miners employed by the American Fuel Company were locked out and told they would be blacklisted for joining a newly established local of the UMWA.[17] Although it was feared that the newly organized local unions in the Utah fields might strike in sympathy with the Sego miners, this did not occur. James Doyle, a former mayor of Victor, Colorado, was employed by the Federal Labor Mediation and Conciliation Board to quietly investigate the charges that the thirty-seven miners were fired and deported out of state because of their union affiliation. Doyle concluded the Sego miners had struck in violation of their own union laws without presenting their grievance to the local union officials. Doyle also noted the men had not been deported.[18]

R.W. Van Derck of the American Fuel Company reported that the altercation arose when a Greek miner, John Muteious, was reprimanded by his foreman for violating safety rules. The miner had drawn a knife on the foreman and afterward was discharged for his conduct. Some thirty-two Greeks protested the action, demanding that their countryman be reinstated and the foreman be fired. When the superintendent refused, the miners walked out in protest.[19]

At the same time the Greek miners were protesting the treatment of John Muteious in Sego, William M. Knerr, industrial commissioner of Utah, telegraphed Secretary of Labor William B. Wilson that according to representatives of the UMWA, coal companies in Carbon County were employing sheriffs to deport UMWA members. Knerr requested that the Department of Labor investigate the allegations.[20]

In January 1919, a confrontation seemed imminent when thirty members of the UMWA in Storrs were discharged by the Spring Canyon Fuel Company. According to John McLennan, when the miners informed Superintendent A. E. Gibson that they intended to form a local union of the UMWA, Gibson replied that if they did, the mine would be closed down and the miners would be forced to leave the company houses. The men were discharged when they consummated the organization of the UMWA local.[21] Writing in the *Utah Labor News*, Henry Sweet described the event in a dramatic manner:

> Down in Spring Canyon, in the name of antagonism to the union, A. E. Gibson, mine superintendent . . . has driven men from their homes and families, refusing them shelter in the company house, where some of them have paid rent and lived for years. He has separated husbands and their wives by illegally and unlawfully evicting men, and he has committed the unspeakable outrage of compelling two families, consisting of women and children, with their men folk, to immediately leave the mining camp. . . . Well knowning that one of these women was about to become a mother, A. E. Gibson brutally ordered the husband, who had long been in the company's employ, to leave their company-owned town. And when the poor fellow made the plea that his wife could not be moved, the inhuman brute of a mine superintendent, A. E. Gibson, ordered up one of the company's wagons, dumped both families into it, and ordered them to "git."
>
> All this transpired during zero weather, the stork following this rough coal wagon for a long, bitterly cold journey. . . . Their only accusation was that of asking the right, guaranteed them by the war labor board, of banding together into peaceful unions. They had been charged with no crime, they had merely joined the mine workers' union, and for this men were taken from their beds at daybreak, without any semblance of law, forced to leave the homes they had known for years, and many of them having wives and small children, were forced to leave their offspring to the charity of their fellow workers, and seek employment far from the scene of the outrage.[22]

J. William Knight, manager of the Spring Canyon Coal Company, denied the charges of discrimination against the miners for affiliating with the union. Instead, he explained that the company's No. 3 mine had been closed down and that fifty men had been discharged because of a great decrease in the demand for coal. Knight instructed Gibson not to discriminate against those who wished to belong to the union. Knight acknowledged that instructions were given for the discharged men to vacate the company houses but noted the men were not to be compelled to move if it would entail any suffering.[23]

Nevertheless, the incident was investigated by the Federal Mediation and Conciliation Office of the Labor Department. When Charles T. Connell, mediator for the Labor Department, threatened to refer the dispute to the United States Fuel Administration unless the issue could be quickly settled, General Manager Knight met with John McLennan and promised that there would be no further opposition to the union and that the discharged men would be reinstated as soon as possible.[24] In his final report on the mediation of the controversy, Charles T. Connell concluded, "The publicity given the Carbon County controversy by the Spring Canyon Company will have a salutory effort upon the other coal companies operating in Carbon County, Utah."[25]

The Spring Canyon incident made little difference, and in March 1919, John McLennan wired Assistant Secretary of Labor H. L. Kerwin that the "Utah Fuel Company [was] discharging men for joining the union and driving them from their own homes which they were encourage[d] to build on company property. . . ."[26] McLennan requested that the charges be investigated and an effort be made to adjust the situation.

Eight months later, the United States Fuel Company in Hiawatha was charged by the UMWA with firing men because of union affiliation. In a letter addressed to Secretary of Labor Wilson, Robert G. Bolton and Jasper Metzger, local organizers for the UMWA, described the discharge and eviction from company houses of Frank Bonacci, his wife and six children; William Gree, his wife and six children; William Beanette, his wife and four children; J. M. Klanger, his wife and child; Angelo Plant, his wife and three children; Tony Dambrosio, his wife and four children; Camillo Garofato, his wife and four children; and Hick Bolder, his wife and two children; plus several single men.[27]

Bolton and Metzger requested that the Department of Labor investigate the situation. However, no action was taken by the Department of Labor against the United States Fuel Company, and the discharged miners were forced to seek employment in other mines. Union members felt that the thirty Hiawatha miners, especially the local union officers, were blacklisted. In reporting the status of the Hiawatha union to John L. Lewis, William Bennett, a Hiawatha miner and UMWA member, noted:

> On Nov. 7th to 12th their was about 30 of our members discharged, this included all local Officials, men without families were given 24 hours to get out of town, and men with families were given 5 days to vacate the town, these orders were enforced by company gards, some of our members went to other mines to work, but we find that the officers of this local cannot get work in any of the mines and all who took an active part in the organization, when they apply for work they are told by the mine Supt. that he is not hiring any men, there are known cases where our men have been refused work and other men hired in fifteen minuets after, we belive this is to be a clear case of discrimination, and further we belive their is what is known as a Gentlemens agrement amoung the coal opperators of this State. We belive that when a man is discharged by a Company there is a letter sent to all Mine Supts in the state giveing names and description and instruction not to hire certin men.[28]

Frank Bonacci, one of the men discharged from Hiawatha because of his union activities, would occupy center stage for much of the next thirty years as the Utah coal union drama unfolded. He was born November 22, 1885, in Decollatura, a town in the province of Catanzaro in the southern Italian region of Calabria. Arriving in the United States in August 1896, he lived for two years in Carbondale, Pennsylvania, before heading west for British Columbia to work on the Canada Pacific Railroad. In 1903, he took a job in a coal mine near Michel, British Columbia, and joined the WFM. Two years later, the local union affiliated with the UMWA; and for the next eight years, Bonacci assumed an active role in the local, filling a number of committee assignments and serving terms as president and as secretary. Bonacci's Canadian stay was interrupted on at least three occasions. In 1907, he traveled to Sunnyside, Utah, to visit his brother Dominic and sister Mrs. Ross Fazio. Earlier, in 1904, he had returned to Italy for a visit and made another trip in 1910 when he married his wife Filomena.

Returning to Canada with his new bride in 1911, Bonacci took up residence in Bellevue, Alberta. After their first child, a girl named Marianna, was born on January 15, 1912, the couple left Canada for Sunnyside and the comfort of family members close at hand. For the next five years, Bonacci worked the Sunnyside mines before moving to Helper in 1917. In Helper, he was employed by the D&RGW and joined the

International Association of Machinists. Bonacci stayed with the railroad only a year. In 1918, he moved to the United States Fuel company town of Hiawatha to mine coal and to anchor the newly organized UMWA local after his election as financial secretary in March 1919. Later in the year, Bonacci's influence expanded with his appointment as subdistrict board member of District 15. Following his dismissal from Hiawatha, he secured work at Kenilworth.[29] The Independent Coal and Coke Company had permitted its Kenilworth miners to organize a local union that met in a hall owned by Dominic Bergera just outside the company property boundary.[30] In Kenilworth, Bonacci was elected financial secretary of the local organization. Although the Kenilworth union did not win recognition for bargaining purposes, it was perhaps the strongest UMWA local in Utah.

By the fall of 1919, the UMWA had organized six local unions in the Utah coal fields: Local No. 4359 in Carbon; No. 4421, Scofield; No. 4422, Kenilworth; No. 4436, Castle Gate; No. 4531, Hiawatha; and No. 4550, Storrs.[31] It was apparent, however, that the coal companies still held the upper hand. The union lacked the strength to strike against the coal companies and could only work through the federal government to insure that men joining the union would not be discharged for affiliation.

During World War I, union coal miners were bound by a wage agreement made in September 1917 that was to run the duration of the war but not beyond April 1, 1920. When armistice was declared in November 1918, coal miners felt the wage agreement could be terminated. Coal operators argued that the agreement was still binding because the war had not been ended by a peace treaty. In September 1919, the UMWA annual convention was held in Cleveland, Ohio. Delegates at the convention approved several demands, which included a nationwide contract, 60 percent wage increase, a six-hour day, and a five-day week. It was decided to end the wartime contract on November 1, 1919, and to call a nationwide coal strike unless the demands were met. Coal operators, however refused to negotiate with the UMWA before April 1, 1920.

President Woodrow Wilson, acting under the influence of Attorney General A. Mitchell Palmer, enjoined the miners from striking two days before the strike was to begin. When Wilson acted, contrary to promises that the law would not be used to interfere with labor activities, organized labor consolidated its support behind the UMWA.

Although UMWA officials technically complied with the restraining order, 394,000 of the nation's coal miners walked out on November 1.

Taking their lead from Attorney General Palmer, the leading spokesman against the "Red Menace," newspapers saw the coal miners' strike as another attempt by communists and radicals to foment revolution in the United States. Attorney General Palmer obtained a permanent injunction ordering the UMWA leaders to end the strike by November 11. John L. Lewis acquiesced and ordered the strike cancelled, declaring, "We are Americans, we cannot fight our government." Nevertheless, many miners continued the strike; and on December 3 Lewis and eighty-three other UMWA officials were cited for violating the injunction. The contempt charges and adverse publicity against the UMWA led to the acceptance of a proposal by President Wilson that offered the miners an immediate 14 percent wage increase and promised that an arbitrating commission would be appointed to review the other demands. By December 10 the strike ended; and three months later, the arbitrating commission recommended a 27 percent wage increase but no change in working hours or conditions. A two-year contract, based on the commission's recommendations, was signed by the miners and coal operators on March 31, 1920.[32]

In Utah, coal operators feared that their miners would join the nationwide strike. The week before the strike, an agent for the coal companies in Spring Canyon secretly recorded the sentiment regarding the strike issue. He attended a meeting at Latuda on October 23 in which the miners declared they had no personal grievances against the coal company but voted in favor of the strike as a means of securing union recognition. Strike sentiment was not so strong in Rains as in Latuda. The strong union supporters, mostly Greeks and Austrians, "were gotten rid of in August. . . . The newcomers, mostly Americans, do not seem to favor a strike, but would probably leave camp if there is one." Men working at Standardville were found, with one or two exceptions, to be union men, and the prospects for joining the strike were high. It was reported that the Italians and other foreigners had smuggled a machine gun into camp to use in case of trouble during the strike. In Storrs, the investigator found little strike sympathy but qualified his report with "but of course you can't always tell."[33] The foreign-born miners in Peerless beat the national strike by five days when they walked out after a Greek miner complained to the superintendent that he was not getting a fair weight. The superintendent cursed the Greek, then fired him. The miners struck immediately, demanding that the offended miner be reinstated.

On October 31 the *Salt Lake Tribune* reported that Utah coal operators were making plans for a probable strike.[34] The morning of

November 1, 100 soldiers from the 21st United States Infantry at Fort Douglas, under the command of Colonel George L. Byram, arrived in Carbon County. Byram, after a long career in the Indian wars, Spanish-American War, and Philippine insurrection, had been called out of retirement at the beginning of World War I to assume command of the prisoner-of-war barracks at Fort Douglas. In explaining why federal troops were sent to Carbon County, District Attorney W. W. Ray declared that the government's intention was to protect any man who wanted to work and to protect the property of the coal companies.[35]

The Fort Douglas troops were reinforced on November 2 by a company of 115 men of the 32nd Infantry from Camp Kearny, California. The federal troops were to insure that shipment of Utah coal not be curtailed because a state of war technically still existed between the United States and Germany, despite almost a year of peace. Special detachments were assigned to watch mine tipples and railroad bridges to prevent strikers from blowing them up.

As the date for the strike approached, local union officials were unsure whether Utah would be included in the nationwide strike call. Their situation during the last week of October was extremely difficult. They had to prepare the Utah miners to strike if so instructed by John L. Lewis, but it was also necessary to use caution in case they received instructions not to join the strike. Utah presented a difficult decision for Lewis. The demonstration of Utah's solidarity with the nationwide strike would be a positive expression for other striking miners, yet a divisive strike in partially organized Utah would jeopardize the substantial progress made during the past year. Finally, on October 31, a few hours before the strike was to begin, John McLennan received a telegram signed by John L. Lewis instructing him to withhold the strike call from Utah.[36]

Immediately, McLennan instructed the local union officers to keep their men at work. However, some miners did not receive the word or chose to strike regardless of the union president's instructions. Reports from Wattis and Latuda indicated the work force was 25 percent of normal; at Standardville, Storrs, and Kenilworth, 50 percent of the miners came to work.[37] On November 1 an Associated Press release from Chicago reported that approximately 1,000 Utah miners were out on strike.[38] The walkouts were short-lived and within a few days the would-be strikers returned to the mines.

In retrospect, Lewis wisely withheld the strike call from Utah. The presence of numerous federal troops, with more on standby, would have greatly curtailed union activity. Also, there was a real question about the actual strength of the union in the Utah coal fields; and with public feelings strongly against the strikers, it was an inopportune time to test that strength.

In the end, the UMWA organization in Utah was indirectly praised for its loyal, patriotic, and conservative action. In Utah papers, the union was not attacked as a harbinger of revolution and bolshevism as it was in other parts of the country. Nevertheless, Utah was not immune to the supposed radical threat. Joseph Granitt and Joseph Benvenute were arrested in Standardville for preaching anarchism, and Wobblies reportedly streamed to the Utah coal fields agitating trouble.[39]

However, the nationwide coal miners' strike proved a misfortune for Utah's union efforts. Those efforts were halted in the face of the coal operators' victory and in the retreat of the Federal Labor Mediation and Conciliation Board, which had acted as a watchdog of anti-union practices. Union leaders lamented that with the 1919 strike, they lost practically all their organization in Utah.

Charges and countercharges leveled by District 15 officers and organizers against each other deterred the successful unionization of the Utah coal fields in 1919. Much of the controversy involved the supervision of the organizational work in Utah by John McLennan. This union "infighting" compromised the effectiveness of the locals.

The first issue centered on the Helper local's assessment payment to the district headquarters. It had been the practice that when an organizing drive or strike was underway in a particular coal field, the local unions were exonerated from paying the usual district and national assessment. Exoneration was granted to the Helper local; but when District 15 Secretary-Treasurer H. C. Stewart wrote John McLennan regarding $200 sent by the district for an investigation of the trouble at Sego in October 1918, McLennan confused the two issues.[40] He accused Stewart of wanting to "jeopardize UMW organization in Utah for the small amount of tax of that one local union." Stewart resented the accusation, and once again asked McLennan for an accounting of the $200 and for the money collected from the Helper local which was to be used for the organizational work in Utah.[41]

A second financial issue involved the Hiawatha local. In December 1919, the District Auditing Committee examined the books of Local No. 4531 and found a shortage of $180 which John McLennan had collected but not accounted for. In its report to John L. Lewis and Secretary-

Treasurer William Green, the committee requested that McLennan turn over the $180 to the local union and show why fifty-one members had been accepted into the Hiawatha local without paying dues.[42]

John McLennan failed to account for the missing funds; and on January 7, 1920, H. C. Stewart formally charged McLennan with misappropriating union funds and with hiring a field worker "who was not fit to serve in that capacity." Elaborating on the second charge in his communication to John L. Lewis, Stewart reported that McLennan had hired an organizer named James Doran and another unnamed individual who roomed "at a house of ill fame." Both men were heavy drinkers and, to the disgrace of the union, often appeared drunk in public.[43]

The charges against McLennan came during the District 15 election in which he was a candidate for district president. In his reply to Lewis, McLennan dismissed the charges as campaign rhetoric.[44]

McLennan won the election for district president; and at a District 15 board meeting held on January 16, 1920, a few days after the election, the lame-duck district officers attempted to pass a resolution declaring the recent election null and void. When this motion failed, another was proposed requesting that an investigating committee be sent from national union headquarters to District 15 and if the committee found it justified, revoke the autonomy of the district. The motion passed much to the disappointment of the newly elected officers who felt the action was taken in a spirit of revenge because of their victory.[45]

Two weeks after the January 16 meeting, a second meeting of the executive board of District 15 was called; charges of malfeasance in office were brought against President George O. Johnson, Vice-President Robert Llewellyn, Secretary-Treasurer H. C. Stewart, and Board Member Frank Bonacci. The prosecutors twisted the earlier resolution, supported by the four men, to imply that through such a resolution, the men were making a direct request that the district's autonomy be assumed by the international organization. Since the action affected the interests of the whole district, the men were charged with malfeasance because they had not followed provisions of Section 2 Article 12 of the district constitution, which required that questions affecting the interest of the entire district be submitted to a referendum vote before the executive board took any definite action.[46]

Secretary-Treasurer Stewart resigned before the charges were filed, and the other three men were found guilty and immediately suspended from office. O.F. Nigro was authorized to assume the duties of acting president until April 1 when John McLennan would assume the position. The office of vice president was allowed to remain vacant until April 1; Felix Pogliano assumed duties as district treasurer for the remainder of Stewart's unexpired term.

Frank Bonacci had been appointed board member from Utah's Subdistrict 7 by President George O. Johnson in December 1919. He was perhaps John McLennan's strongest critic in Utah and, as secretary-treasurer of the Hiawatha local, was a principal character in the audit of the Hiawatha books, which resulted in charges of misappropriation of union funds against McLennan. In a lengthy letter to John L. Lewis five days after his suspension, Bonacci gave an illuminating picture of union affairs in District 15, particularly in Utah.

> By request of my many friends Dist President Mr. Geo O. Johnson appointed me Board member of Sub. dist No. 7 Utah and was approved by the board Dec. 9, 1919. Being out of work it was a little help to keep my big family of a wife and six children, but I was not the kind of a man Mr. John McLennan wanted to represent the men of Utah at the board meetings and since that time has kept a steady fight on me and got his right hand man Int. organizer James Doran to go around the locals asking local unions to pass resolutions asking my removal and appoint James Doran. Also charging me as a spy of the U.S. Fuel company of Hiawatha, Utah where I was working before my discharge Nov. 7 and that I am responsible for the discharge of those men, including my discharge as a fake.
>
> This is done in an effort to discredit me before the men in Utah. For haven sake Mr. Lewis, wat kind of a union those people are trying to organize in Utah. It seams to me they want to make a little state of Germany out of Utah and a union man has got to be *Blind Deaf and Dumb.*
>
> Some radical changes has got to be made out here to save the situation if a union is going to be established in Utah and Colorado. . . .
>
> I am more then ever convinced that the only salvation of this dist is for the International organization to step in and clean the whole bunch from office for the time being and when in their oppinion to give the membership another clean election because as long as we have a bunch of office seakers knocking one another, stealing votes, changing election returns, making trips from one camp to another at the expenses of the national for their elections, having local officers to instruct the men who to vote for and mark ballots for those not present, and putting the election's campaign ahead of anything else, I do not believe that the affairs of the dist will ever prosper to any success.[47]

The Bonacci-McLennan controversy was reminiscent of earlier friction between idealistic foreign-born organizers and their English-speaking co-workers. In 1904, Charles DeMolli was hailed by his countrymen as greater than George Washington, and the Italian support of the 1903–4 strike was as much an expression of loyalty to DeMolli as it was a demonstration of the Italian miners' acceptance of unionism. When DeMolli was forced out as an organizer for the UMWA because of personality clashes with American union leaders, the rank and file in Utah speculated what the union would be if DeMolli, not John Mitchell, were president and rejoiced in the unfounded rumors that DeMolli was going to start his own union.

Louis Tikas, a Greek organizer in the southern Colorado coal fields, experienced a similar frustration in working with American union leaders during the hard-fought Colorado strike of 1913–14. Two months before his murder by Colorado militiamen and ten days after his dismissal as a union organizer, Tikas wrote to John L. Lewis giving an account of his fifteen-months' employment by the union. Tikas blamed national organizers who, because of their petty jealousy and apparent dislike for the Greek strikers, sought to undermine his influence with the strikers.

> These men, seated around a large table in comfortable chairs in the Trinidad office, smoking their good fifteen cent cigars, riding in automobiles at the expense of the organization and their work limited to the making of a few speeches, presumably thought that I desired their soft . . . [?] . . . and that would oust them from their jobs, so they . . . plotted to discredit me with the people. . . .
>
> They would first of all displant my name from any mention in the newspapers and secondly and by far the most important, for I care little about newspaper notoriety, refuse to recognize my orders for supplies from headquarters. They knew my power over the people and knew that the quickest and surest means to discredit me with them, was to show the people that I no longer had influence with those in authority and that when in want and in crying need of the necessities of life, they must look elsewhere to be supplied.[48]

John L. Lewis requested that an investigation be made of Bonacci's dismissal and of the charges against the three district officers. The District 15 committee reported to Lewis, "We find that Frank Bonacci, was put on as a subdistrict Board member, by the District Board and discharged by said Board, and that the International Union has no jurisdictions on this case."[49] The decision reaffirmed the principal of district autonomy, which Lewis, in this case, was willing to accept. Nevertheless, the antagonism between foreign-born organizers, like DeMolli, Tikas, and Bonacci, and English-speaking leaders, like McLennan, would hinder the union effort until foreign suspicions and nativistic prejudices abated. Ultimately, the shared need for unionization worked to ease those tensions over time.

The dissastisfaction with District 15 officers led the Kenilworth local to petition John L. Lewis that Utah be transferred from Colorado's District 15 to District 22, which encompassed the Wyoming coal fields. The resolution, adopted at a meeting on June 21, 1920, repeated the complaints against John McLennan and concluded that the support needed by the Utah union could come only from Wyoming.[50] Frank Bonacci, as secretary-treasurer of the Kenilworth union, undoubtedly played a major role in the decision to transfer.

No action was immediately taken on the request. In February 1921, the Scofield local expressed concern with the manner in which Utah had been abandoned by the union. In a letter addressed to John L. Lewis requesting assistance, the local sounded a deep resentment against the present arrangement.

> Utah, at one time, was fairly well organized and through the representatives of the International Organization we went to *Hell*. The spirit of the miners, without anymore assistance, is getting to be worse than ever, without *no one* to represent us in our struggles which we have daily.
>
> Now, we do not believe that only the men from the "East" are capable of representing us, but, we have men in Utah, [Frank Bonacci] competent to do this work and that have more confidence in the Utah miners than those that come from the East.
>
> The men sent from the East need at least three months to get acquainted with the conditions in this state. We have not had many that *did* stay three months in Utah.
>
> We, the members of Local Union #4421 uphold the strongest local in this state, and we are making a demand, to the Executive Officers of the International Union, that immediate action shall be taken and organizers sent to this state, because we are losing members and do not want to lose what members we have.
>
> After, if Executive Officers do not take immediate action on our demands, we will publish this in the Journal, [*Journal of the United Mine Workers of America*], so that the mine workers throughout the country can see how we are being treated.

We want to have a definite answer to this, to know what steps you are taking by Feb. 19, 1921.

Respectfully yours, B. Mroosh, B. Williams, T. Mulligan, C. Mancuso, C. Zavis. Committee on Correspondence, By And. L. Aho, Rec. Secy. L.U. #4421, UMW of A.[51]

Although John L. Lewis did not meet the February 19 deadline, he did consider the Utah question. Wyoming officials were contacted to determine their willingness to take on responsibility for Utah. They agreed to accept the liability until the Utah fields were sufficiently strong enough to have their own district. The transfer was accomplished on July 1, 1921.

In reporting the action to the International Executive Committee, John L. Lewis noted the union had lost considerable ground in Utah since the 1919 strike, despite efforts by District 15 officials to change the situation. Lewis stated that District 22 officials were particularly interested in the Utah situation since coal companies operating in Wyoming were also found in Utah. With the Utah miners' petition as a catalyst, a joint agreement was made with District 15 and District 22 officials clearing the way for the change in jurisdiction.[52]

Reports are not available indicating the number of Utah miners who actually belonged to the UMWA at the time of the transfer to District 22. A year earlier the semiannual report of the secretary-treasurer of District 15 reported Utah membership as follows:[53]

Local Union 4245 Sego	386
Local Union 4335 Latuda	305
Local Union 4339 Wattis	329
Local Union 4345 Rains	14
Local Union 4348 Standardville	48
Local Union 4421 Scofield	580
Local Union 4422 Kenilworth	273
Local Union 4436 Castle Gate	129
Total:	2,064

The entire District 15 membership numbered slightly less than 11,000. It is doubtful that by July 1, 1921, the union membership in Utah remained over 2,000. However, with the transfer to District 22, the Utah union men were confident that the new arrangement would lead to an increase in membership.

Chapter VII

The 1922 Strike

The American coal industry did not prosper during the 1920s. New mines, opened to meet the demands of World War I, glutted the market as the peacetime need declined. Competition was intense, and coal production shifted from union mines in Pennsylvania, Illinois, and Ohio to the nonunion mines in western Virginia, Kentucky, Tennessee, and Alabama. Wages were reduced in the nonunion mines; and although wildcat strikes occurred and attempts were made to organize the workers in these areas, John L. Lewis insisted that men under contract honor their agreements and refrain from any sympathetic or protest strikes.

Nevertheless, with contracts due to expire on April 1, 1922, the UMWA began to prepare for another strike. Negotiations held early in 1922 between UMWA officials and coal operators broke down when the operators insisted that wages would have to be reduced in order to compete with the nonunion coal. As a result, the coal industry edged toward conflict as the contract period drew to an end.

Over 500,000 bituminous miners and 150,000 anthracite miners responded to the strike that began on April 1, 1922. The strike, the largest in American history to that time, had two objectives: the organization of nonunion mines and the prevention of the wage cuts which the operators deemed essential.

The Intermountain West followed the national trend as the union coal of Wyoming competed with the nonunion coal of Utah, and miners in both states faced wage cuts of nearly 30 percent.

On July 1, 1921, Utah, which had been a part of District 15 of Colorado since 1903, became part of Wyoming's District 22. Immediately after the change of districts occurred, union officials from Wyoming began arriving in Carbon County. In an open letter to Utah coal miners published in the *Helper Times* and the Price newspaper the *Sun*, William Houston, intermountain organizer for the UMWA, boldly announced the change in status and a plan for a major effort to organize Utah, directed from the newly established union headquarters in the Helper Hotel.[1]

Almost immediately, organizers succeeded in bringing new life to the six Utah locals that had been dormant since the end of 1919. This rebirth was reflected in the *Wyoming Labor Journal*, which on August 21, 1921, published the results of a recent election of officers for Local No. 4422 in Kenilworth. Reports also came from the Utah fields that "since being taken into District No. 22, a renewal of interest is being manifested by the miners."[2] During the winter of 1921–22, the organizing work in Utah pushed ahead under the direction of Martin Cahill, president of District 22, James Morgan, secretary-treasurer, and William Houston and Tony Radlaj, organizers.

Nevertheless, with a nationwide coal strike approaching, the strength of the UMWA in Utah still seemed in question. The *Sun* cautiously predicted that the strike would not have any great effect in Utah since the mines were not organized and since the Utah miners had not participated in the November 1919 nationwide strike. The Price newspaper cautioned that because the effectiveness of the newly arrived union organizers could not be measured, even "the operators hesitate to make any clear cut prediction as to just what will occur."[3]

If the Utah coal operators were unsure about the union strength in the Utah coal fields, so were UMWA officials. When the nationwide strike was called for April 1, Utah was not included in the strike order. However, when a reduced wage scale was put into effect in Utah on April 1, most of the miners from the Spring Canyon mines walked out. Workers from Clear Creek, Winter Quarters, Scofield, Kenilworth, Castle Gate, and Rolapp soon followed. With this local action, it was obvious that Utah's miners were determined to forge the link that would bind Utah to the UMWA; the district organization was requested to assist the striking miners and complete the organization of the Utah fields.[4]

The reduced wage scale—the initial reason for the Utah strike—represented a loss of approximately 30 percent of miner income. Wages for the day laborers dropped from $7.95 to $5.25 per day, and rates for mined coal fell from seventy-nine cents to fifty-five cents per ton.[5] In addition, such long-standing grievances as the coal companies' cheating on the weighing scales, high rents for company housing, favoritism, and other abuses became issues in the strike. By the fourth week of the strike, approximately 70 percent of the work force was out on strike.[6]

Operators responded in a variety of ways to the initial threat of a strike and, later, to the strike itself. In an effort to ease the impact of wage reductions, they made modest reductions in the cost of rent for company houses, reduced the cost of coal to miners, and decreased the charge for mining supplies such as powder and fuses.[7] However, prices for coal to consumers remained unchanged, and the 15 percent reduction in housing, coal, and supplies did not begin to compensate for the 30 percent wage cut.

Mine owners also publicly denied that there was any discrimination against union men, maintaining that "the question of union membership enters into the question of employment no more than does the man's religious belief."[8] Later, after the strike was in full swing, the coal operators reversed their position declaring that they would "never reemploy any of the men who had fomented and perpetuated the present difficulty."[9]

The operators also announced that coal production was continuing to meet market demand; and that while many were out on strike, those who wanted to work were getting five days a week instead of three, as was the case before the strike. Union officials countered that even though the coal companies were continuing to ship coal, 80 percent came from supplies stored before the strike began.[10]

With the threat of a strike, mine guard ranks were increased to keep careful watch on company property. Once the strike began, they maintained deadlines along the edge of public lands as well. Guards were a point of friction. They were blamed for being "domineering," and it was alleged that "their general attitude was of a character which would provoke trouble any place in the United States."[11] Utah, the *Wyoming Labor Journal* lamented, would rank with the "feudal dominions owned by the coal operators of West Virginia, Colorado and . . . Pennsylvania" until the state abolished the private armies of Utah's coal operators. The Pleasant Valley Coal Company was proclaimed the heirs of the "cossakism" of Tsar Nicholas before the Russian Revolution.[12]

Trouble often erupted when guards carried out evictions. Unmarried miners who lived in company boarding houses were, in some cases, evicted without any chance to pack their belongings. The occupants of company houses were given between five and ten days to move. Those who did not leave voluntarily had their belongings hauled off company property and dumped in the snow. Strikers were prodded along with the muzzles of

Winchester rifles. Businessmen who had built stores on leased coal company property and who sympathized with the striking miners found themselves in the same situation as the evicted strikers. George Zoumadakis and two of his Greek countrymen, who built a large store and coffeehouse on land under a ninety-nine-year lease from the coal company, were forced to close the store despite a lease provision that there would be no interruption of service. With food spoiling and no immediate prospect for reopening the store, the businessmen sold their $18,000 inventory to the coal company for a tenth its value.[13] In addition to the guards, spies employed by the coal companies infiltrated the miners' union and kept close tabs on the activities of union leaders.[14]

The disruptive effect of the 30 percent wage reduction was ignored by company spokesmen who argued that the Utah strike was being led by outside agitators for the sole purpose of aiding the nationwide strike and not for the benefit of the Utah miners.[15] More specifically, the pro-company Price newspaper the *Sun* labeled the Carbon County strike a "Wyoming strike."[16] The *News Advocate*, also published in Price, espoused nativistic sentiment by suggesting that union officials were promoting violence by arming the foreign element with a lot of "expensive ugly automatics."[17] When a large percentage of the foreign-born miners, especially the Greeks, joined the strike, the local papers branded them a "bunch of vicious bulls" who had no desire to become "Americanized" and whose low morals and lawlessness were a curse to the country.[18] On the other hand, the *Wyoming Labor Journal* concluded that the coal operators considered immigrants acceptable "as long as they uncomplainingly pile up wealth for their employers, but when they tire of oppression and undertake with others to obtain justice the companies begin a campaign of persecution and anti-foreigner agitation."[19] Nevertheless, the attacks on foreign-born miners swayed public opinion and greatly aided the recruiting of strikebreakers among the state's rural population.

Spokesmen for the coal operators declared that the efforts of the union were directed first toward the immigrant groups and explained that the organizers succeeded because of the secret societies among the foreign-born miners. The operators charged that the union organizers had only to "win over the leaders of any one group in order to bring into their fold that entire national group."[20]

Company officials showed a great reluctance to deal with the union leaders out of fear that any form of recognition would be detrimental to their cause. On April 15 union representative William Houston offered to settle the strike at the reduced wage scale if the operators would agree to dismiss the armed guards patrolling the company properties; to permit the union men and organizers to hold meetings; to establish an absolutely open-shop employment system with no discrimination between union and nonunion men; and to apply the reduced scale until a wage agreement was adopted for Wyoming, which would then become the basis for a new Utah scale.[21] The operators replied they were not notified of the union offer and refused comment on the proposal. They concentrated on persuading the men to return to work and, when this failed, imported strikebreakers under the protection of company guards to operate the mines. This practice precipitated the first violence in the strike—an incident that occurred when the Utah strike was four weeks old.

In April, a tense situation developed in Scofield, especially after strikers in the company town of Winter Quarters, evicted from their homes and boarding houses, congregated two miles east of Winter Quarters in the noncompany community of Scofield. N. J. Salyards, a post office inspector from Provo, observed, "The feeling was so tense that at the least unusual noise both miners and mine guards would look to their guns." Salyards explained that "anyone coming into the towns is accosted by representatives of the mining companies and miners as to his business and if he thinks he will be long in that vicinity."[22]

The focal point for potential conflict between the strikers and company guards was the Scofield railroad station. A large congregation of strikers, ready to discourage any potential strikebreakers from entering the area, met each incoming train. In defiance of the strikers, officials dispatched a wagon, protected by a mounted and armed squad of guards, to meet incoming workers and escort them to the Winter Quarters mines. On April 27 at 1:30 p.m., as the detachment of company guards left the railroad station without any strikebreakers (who, according to rumor, had been due to arrive on the afternoon train), shots were exchanged between strikers and company guards. Three men were wounded and one deputy's horse was killed. Sam Dorrity, a mine guard and former chief deputy United States marshal, was shot in the leg; Fred Jarvis, a striker, was shot in the back with the bullet passing through his lungs; and a striker, Mike Makesmrticos, was shot through the right shoulder. The company version, as reported by Sheriff T. F. Kelter in his request for the National Guard to be sent to Carbon County, placed the blame squarely on the strikers, charging they had thrown stones at the guards and that a Greek striker,

George Manousos, had fired the first shot. The strikers' version of the incident predictably blamed Dorrity as the aggressor.[23]

When the Dorrity affair occurred, Governor Charles R. Mabey was in California, and Sheriff Kelter's request for the National Guard was sent to Acting Governor H. E. Crockett and Adjutant General W. G. Williams. Kelter declared: "It is a reign of intimidation. They [the strikers] have threatened to kill any man, woman and child who goes to Scofield from the camps that are working, and they have the men and the guns and the inclination to carry out their threat."[24]

County Commissioner A. E. Gibson reported to Acting Governor Crocket that "a large body of foreigners fully armed for a pitched battle are reported to be marching on one of the mines bent for destruction."[25] Despite newspaper headlines that troops would immediately entrain for the Carbon County coal fields, Crockett waited until Mabey returned before committing troops to the strike area.

Anxious to prevent the National Guard from being sent, union officials declared that peace would be maintained and instructed their men to "preserve order under all circumstances and to see that no further trouble occurs no matter what the provocation might be."[26] Taking a much more militant position, M. P. Bales, president of the Utah Federation of Labor, threatened that a general strike would be called if state troops were sent to Carbon County.[27]

Governor Mabey returned from California the afternoon of April 29, two days after the shooting. That evening coal operators met with the governor for three hours in an effort to persuade him that troops were needed. Their position was supported by Arthur W. Horsley, president of the Carbon Stake of the Church of Jesus Christ of Latter-day Saints. However, relying on reports from William M. Knerr of the State Industrial Commission, backed by the pleas of attorney Samuel A. King, Mabey decided that no troops would be sent unless conditions changed. In a meeting with Carbon County commissioners, Mabey placed the burden of law enforcement squarely on their shoulders. "I want you commissioners to get busy. I want you to see that those camps are properly policed and that the lives of innocent persons are protected. Police every camp in the county."[28]

To meet the policing need, forty-nine men were hired from Salt Lake City, Bountiful, and Provo. The new deputies were immediately transported by a special train provided by the D&RGW to Carbon County. The men, paid by Carbon County, were told to "keep their heads" and show favor neither to the coal interests nor the strikers.[29] The strikers in Scofield were not enthusiastic about the new county deputies, who were housed in the bunkhouse of the old Union Pacific mine. The miners protested the action by briefly boycotting a meeting called by Mabey's representative, William M. Knerr. When the meeting finally began, the strikers refused A. E. Gibson, chairman of the Carbon County Commission, permission to speak because he had favored calling out the National Guard and was well known for anti-union activities in 1919 as superintendent of the Spring Canyon mine.

Although the focus of attention during the first month of the strike was on the Scofield-Winter Quarters area, coal miners in other camps also left the mines. Tent colonies were established just outside company property at Hiawatha, Kenilworth, and Sunnyside. Here striker harassment continued as company guards disturbed the tent colonies with their shouting, cursing, and shooting. At Kenilworth, where strikers set up a picket station just off the company property, in a building owned by Dominic Bergera, Superintendent Elwood surrounded the camp with high-voltage barbed wire, closed the road into camp, set up searchlights to scan the canyon, cut the waterline to the strikers' houses, and (with the protection of the company guards) appeared frequently at the outer gate to brand the pickets "cowards and pimps."[30]

Helper, cherished by the strikers as "an oasis of justice in a land otherwise controlled by the operators and their satellites," served as strike headquarters for the Castle Gate, Kenilworth, and Spring Canyon strikers; and Scofield was headquarters for miners from Winter Quarters and Clear Creek. In both communities, tents were set up on vacant lots and on property belonging to sympathizers. Several tents were erected in Price to house the strikers, however, it was well known that the "'official' sentiment at Price has not been very kindly to the strikers."[31]

Despite worry about providing food for their families and concern regarding the success of the strike, life for the strikers was not totally grim. Baseball games were played, usually between the single and married men. Pitching quoits and horseshoes were popular activities. The strikers in Scofield responded to a request by Mayor Lars Jensen to aid in a general town cleanup. When fire broke out there in the Beddoes Hotel, strikers became volunteer firemen and quickly extinguished the flames. A strikers' committee for law and order was appointed in the town; one of its first acts was to declare that bootlegging would not be tolerated and that all prohibition violators would be instantly arrested. In an effort to prevent strikers gathering at the railroad station and the possible recurrence of an event similar to the Dorrity affair, the strikers' committee was appointed to meet incoming trains and inform would-be strikebreakers of the situation. To discourage strikebreakers further, a delegation of ten strikers, with appropriate banners, was posted at Colton where the railroad to Scofield left the main branch of the D&RGW.[32]

During this same period, union officials mounted a campaign to strengthen union ranks and to win support and sympathy for their cause. They accused Utah coal operators of using West Virginia tactics, employing guards in a despotic effort to control strikers. The company gunmen were charged with provoking violence to preserve their soft and profitable jobs. In contrast, the strikers were praised for their stoicism.

In the Emery County communities of Castle Dale and Huntington men were being recruited to work in the mines; organizer William Houston made a special trip to hold meetings. Other organizers visited nearby counties on similar missions.[33] Efforts to discourage strikebreakers from the towns and cities of Utah were reinforced by union attention to the Japanese workers, who, for the most part, had remained at work. Only a handful of the few Japanese in the mines joined the UMWA. Among them was Hideo Kazuta, who had previously been employed at Kenilworth as a team driver. An articulate spokesman for the union interests among the Japanese, Kazuta, had served as an infantryman with the 91st Division in World War I. He spoke and wrote English better than many of the local nativistic detractors of the foreign-born and was as loyal to the union movement as he was to the country he had served during the war. In a letter directed to the general public and to his Japanese countrymen, Kazuta urged that they understand the reasons for the strike and support the union cause.

> As I love America, I love the state of Utah and especially Carbon County. . . . I stand against lower wages in order to save Kenilworth where I was working . . . from utter misery, and to protect the businessmen and farmers of Carbon and Emory counties where the worker's wages is the barometer of prosperity. Every genuine American I feel is opposed to the present tendency to change our standard of living in our beloved country by lowering wages, to that of other lands where kings and a few followers have the lion's share of the nation's wealth and happiness, while the great mass of people face suffering that comes from no work or low wages.[34]

Articulate speakers simultaneously inspired and threatened strikers. The names of many who worked during the strike were listed in the

Wyoming Labor Journal, and in some cases, those men who joined the union and returned to work before the strike ended were formally expelled from the locals.[35]

Despite the efforts of union leaders to maintain solidarity among the strikers, the striking community was most effectively united by the shooting of a handsome young Greek striker. The shooting of John Tenas by Deputy Lorenzo H. Young, like the Sam Dorrity affair, is clouded with accusations of guilt on both sides. Both accounts agree that Tenas was shot at 12:30 p.m., Sunday, May 14, 1922, just west of Helper close to the Ambrosia orchard, near a steel railroad bridge in Spring Canyon. However, from that point on discrepancies begin. According to Young's story, he was en route from Latuda to his home in Huntington. When his automobile became disabled, he took it to a garage in Helper. After strikers threatened him there, he was escorted out of town by the city marshal and began the return to Latuda on foot. While Young was walking along the public highway, Tenas, coming from his Ambrosia home, approached Young from a lane that intersected the Spring Canyon highway. According to Young, when they were about forty feet apart, Tenas pointed a rifle at him and told him to go back. Young immediately drew his revolver. Tenas fired, hitting him in the thigh near the hip; Young returned the fire, felling Tenas with the first shot and killing him with a second when the striker started to get up. Contrary to Young's version, witnesses to the shooting declared that Tenas was unarmed and that when Young left the scene he was not wounded. Witnesses also testified that Tenas was running away from Young when the deputy shot him in the back. The controversy was not resolved. After the Greek's body was examined, a coal company doctor held that Tenas was shot in the front while a Helper doctor maintained the deadly bullet entered his back. No rifle belonging to Tenas was produced, and union spokesmen blamed Young for an unprovoked attack. Young was arrested on charges of second-degree murder but released on $5,000 bail. The case did not come to trial.[36]

The Tenas shooting precipitated another crisis in Carbon County, and many felt the strikers would seek retribution through violence. Headlines once again declared that a "Race War in the Coal Fields is Imminent,"and Carbon County commissioners frantically requested 500 rifles, 10 machine guns, and 10,000 rounds of ammunition from state officials, declaring that "armed warfare between the American and foreign elements hangs on a hair trigger."[37] Tenas's funeral in the Price Greek Orthodox Church and his burial in the Price cemetery were impressive demonstrations of unity among the Greeks. The casket was escorted from the church to the cemetery by flag bearers carrying the American and Greek flags. With the Price band playing a solemn march, 700 mourners, many holding small blue and white Greek flags and lamenting that one so young would be buried so far from his native land because of such a despicable act, accompanied their martyr north toward the cemetery.

On May 15, the day after the Tenas shooting, Governor Mabey journeyed to the coal fields to assess the situation. Mabey chose to hide his identity in an effort to gain a clearer understanding. According to newspaper accounts, he walked six miles through Price Canyon to reach the coal fields. The next morning, when he tried to visit the company town of Kenilworth, a company guard stopped him and refused him entrance when the governor declined giving his name and stating his business. Mabey spent three days in Carbon County. Criticized by many for once again refusing to send in the National Guard, Mabey insisted that the burden of law enforcement rested with local authorities.[38]

The governor did take the lead in drafting and securing acceptance of a five-point peace plan by coal company, union, and county officials. The plan called for aliens to disarm by 6 p.m., Monday, May 22; mine guards to be discharged and their status as deputy sheriffs terminated—the companies were to have watchmen only; remaining deputies to be supervised by the sheriff and be paid by the county, not the coal company; intimidation by both coal companies and strikers to be prohibited with every man permitted to work where and when he pleased; and the state to not interfere until county officers made a demand for citizens to preserve the peace.[39]

Acceptance of the plan by the three factions was at best lukewarm. The coal operators attached reservations that the strikers could not in good conscience accept. The reservations included that the strikers refrain from demonstrating or picketing near company property; that the company watchmen be permitted to carry arms while on company property; that the watchmen have the right to say who would be permitted on company property; and that the deputies under the jurisdiction of the county sheriff be permitted to be hired as coal company watchmen.[40]

Acceptance of the reservations would have, in effect, nullified the plan by merely maintaining the status quo. The UMWA acceptance of the plan was limited to Samuel King's advising William Houston that the union members should adhere "to the letter" of the governor's plan. The

Carbon County commissioners agreed to the plan only after the state loaned them rifles, machine guns, and ammunition.[41]

In less than a week, the peace plan failed. First, the striking aliens refused to turn over their weapons, fearful that they would be defenseless against the armed company guards. Second, county officials, worried about the impact on county finances, dismissed the county deputies who were to enforce Mabey's peace plan. In doing so, the commissioners declared their inability to deal with the situation and insisted that the responsibility for law enforcement rested with Governor Mabey.[42]

The dismissal of deputies was followed by a concerted effort on the part of striking miners to discourage working miners and strikebreakers. They sent threatening letters to men still employed in the mines, and stopped and searched the stages to Hiawatha and Sunnyside for "scabs." At Sunnyside, it was reported that some of the strikers cursed and spat in the faces of the stage passengers.[43] Six Greeks were arrested for holding up the Hiawatha stage.

The stage holdups were followed by the attacking of "man trips" (trains of cars carrying men into or out of the mines). The first incident occurred at Kenilworth on June 2 when a reported 50 to 100 shots were fired at the man trip as it left the Kenilworth No. 2 mine between 4:00 and 4:30 p.m. Fortunately, none of the miners were injured; however, the cars in which they were riding were struck in many places. The strikers blamed the shooting on company guards. A week later on June 9, three snipers fired twenty shots as men left the man trip at the hoist house in Standardville. As at Kenilworth, none of the miners were injured, but one miner's hat was pierced, and bullets passed through the dinner buckets of two others.

Troop F and Battery A of the National Guard were mobilized following the Kenilworth incident and held in readiness for four days. Despite the pleas of the antistrike element, Mabey once again followed the advice of his agents on the scene and refused to order the guardsmen into the coal fields.[44] The situation remained tense with acts of intimidation reported by both sides. Strikers were accused of setting fire to an engine house of the Utah Railway Company in Martin and to two company houses in Sunnyside after they were vacated by strikers.[45] Coal company officials trained powerful spotlights on the strikers' camp in Hiawatha, and "everything was done that could be devised to vex and annoy these . . . men and their families."[46]

As the strike reached a stalemate, the front line of conflict was the dangerous and emotional ground of strikebreaking. The *Sun* viewed it as a fight between strikebreakers—those who worked during the strike, and strikemakers—those who were from outside the state and were bent on making "a desert of Carbon County."[47] The strikers remained adamant in their attempts to prevent strikebreakers from entering the coal camps. Under these circumstances, the next act of violence—one which finally brought the National Guard to Carbon County—occurred.

On June 13 approximately thirty strikebreakers, recruited in Colorado, were to be taken from Castle Gate to Standardville. When the Utah Railway crew assigned to the train refused to handle the engine and single passenger car as a gesture of sympathy with the strikers, coal company and railroad officials resolved to operate the train themselves. Strikers at the tent colony, located at the mouth of Spring Canyon in the area called New Helper, had been meeting the trains transporting strikebreakers at Jacob's Switch where the trains were required to stop. Through this practice, the strikers had successfully persuaded many would-be scabs from entering the mines. In addition, several minor clashes between strikers on picket duty and company guards had occurred at Jacob's Switch. F. C. Henness, superintendent of the Standardville mine, was well aware of the effectiveness of the pickets at Jacob's Switch and, according to the strikers' version, was determined to stop it "either in a peaceful or violent manner."[48]

Thus challenged, the strikers met the next train of strikebreakers at Jacob's Switch on June 14 with a special detachment of pickets. As the train—engineered by C. L. Vaughn, a railroad official, assisted by Arthur P. Webb, a company guard serving as fireman—emerged from Tunnel No. 1 near the mouth of Spring Canyon, shots were exchanged between the train crew and passengers and the strikers. Arthur Webb was killed, and company men H. E. Lewis and William F. Abbott suffered knee wounds. A striker, Andreas Zulakis, was shot in the left forearm. Once again each side blamed the other. The strikers admitted that they were heavily armed but argued that the shots had first come from the train where both guards and strikebreakers had opened fire. The coal company declared it was a deliberate ambush, since shots had come from two directions and the location was ideal for an ambush.[49]

In retaliation for Webb's shooting, a group of men with masks and smudged faces entered the Liberty Pool Hall in Standardville and forced six Greeks to march down the canyon towards Helper. On the way, they stopped at a bakery operated by George Marcoulis; they forced him and two others to join the prisoners. As Marcoulis later testified, he was asleep in a company house across from the bakery in Standardville when five masked men armed with rifles burst into the house, shoved their weapons in his chest, and searched the room. Marcoulis and his two sleeping companions were forced out of the house where they joined the other Greeks. The frightened men were insulted, abused, nudged with rifles, and threatened with execution by hanging or shooting. Finally, the Greeks were told to start running. With bullets whistling around their ears and exploding dust at their feet, the men vanished into the darkness and spent the night walking to Helper.[50]

As the John Tenas funeral served to unite the strikers, so the Arthur Webb funeral had much the same effect on those who maintained that a state of lawlessness had prevailed since the strike began. At the Webb funeral, the Reverend R. C. Jones mirrored strong antagonism toward the union organizers and the foreign-born; he branded them responsible for the lawless acts and demanded they be driven out of the county.[51]

Governor Mabey acted quickly, and at 2:10 p.m. on June 14, the day of the Webb shooting, he ordered the National Guard to proceed to Carbon County. Eleven hours later, at 1:10 a.m. on June 15, detachments from Salt Lake City, Brigham City, and Ogden were en route to the strike area. Along with his National Guard mobilization order, Mabey established martial law in several Carbon County towns and coal camps.[52]

Headquarters for the National Guard was established in the Utah Hotel at Helper with Major Elmer Johnson commanding 13 officers and 152 enlisted men. The remainder of the Guard, 8 officers and 53 enlisted men, was assigned to Scofield, serving under Major Hamilton Gardner until he was relieved on July 13 by Captain James B. Tucker.[53] The National Guard headquarters in Scofield was established at the old Union Pacific Hotel.

With little time for rest, guardsmen at Helper began rounding up strikers on the morning of June 15. Approximately 300 miners were herded to a vacant lot west of Helper where Major Johnson, aided by a Greek interpreter and a chairman of the miners' picketing committee, read the governor's proclamation, explained the meaning of martial law, and asked the strikers to turn in their weapons. Approximately twenty-five men complied. After a search of the entire group, an additional fifteen or twenty weapons were found. The strikers then were allowed to leave, with the exception of twenty men accused of participating in the Webb shooting.

Their names had been given to Guard officers by H. E. Lewis of the Standard Coal Company.[54]

The National Guard quickly made its presence known by posting sentries on all roads leading to and from Helper and by placing machine guns on the hills above the strikers' camp at New Helper. Within the Helper city limits, several three-man patrols were dispatched to prevent street meetings and other gatherings and to enforce a 10:00 p.m. curfew. The patrols stopped all incoming vehicles and pedestrians and searched for weapons and prevented those who failed to secure "proper authority" from entering Spring Canyon. When citizens refused to turn in their weapons voluntarily, the Guard conducted numerous searches. Additional weapons were uncovered, often by digging in soft spots in the ground.[55]

The initial policy of the National Guard regarding the strikebreaker issue favored the strikers. Major Elmer Johnson declared that under martial law the coal operators would not be permitted to ship strikebreakers into the field. Strikebreakers en route to the mines at Castle Gate, Rains, and Hiawatha were turned back by guardsmen. At Clear Creek, Major Elbert D. Thomas, who had fulfilled an LDS mission to Japan and who would later serve from 1933 to 1951 as United States Senator from Utah, uncovered five Japanese strikebreakers by questioning them in their native tongue.[56]

UMWA officials praised the National Guard for its neutral position. In a letter published in the *Wyoming Labor Journal*, union organizer John Ramsay noted:

> Judging from the wail of the local press, Carbon county, Utah, coal operators are not so sure of their position as they were when they finally induced Governor Mabey to order state troops into this field. The advent of the militia meant to them, as it has in most previous strikes, that they could freely import strikebreakers under any form of misrepresentation and be protected in so doing by the militia, but to their bitter chagrin they are now discovering that the National Guard of Utah is not a strikebreaking agency. It protects men who are voluntarily seeking employment at the coal camps, if necessary, but will not permit men to be shipped in under false representations as to conditions as the practice has been since the strike started. Operators are therefore raising a howl and are as anxious to get the troops recalled as they were primarily to have them in field, but without avail so far, and a condition we hope will continue.[57]

Nevertheless, while Ramsay's letter was being published by the Cheyenne paper, the National Guard stopped preventing the strikebreakers entry and pursued a policy in harmony with the coal operators. The headline of a *Salt Lake Telegram* article of July 6, 1922, declared, the "National Guard is not Preventing Men from Replacing Strikers." The article reported that because of a potential coal shortage that would bring great hardship to the public, the National Guard would no longer prevent workers from entering the coal fields. Adjutant General W. G. Williams denied that a change of policy had occurred but confirmed that strikebreakers would now be allowed in the mines.

Another blow was struck at the union efforts when Guard officers, using their power under martial law, denied union officials the right to address gatherings of miners who were at work in the coal camps. Meeting with Governor Mabey and Adjutant General Williams, attorney Samuel King unsuccessfully argued that if nonunion miners were allowed to work under the protection of the National Guard, the strikers should be guaranteed the right to picket and the right of free speech.[58] Voicing his resentment against the change in the National Guard's policy, King contradicted the earlier optimism voiced by organizer John Ramsay.

> True, at the outset, the operators were forbidden to import strike breakers, and protection was given individuals seeking employment on their own initiative. This policy received the hearty endorsement of the union officials. It was just and fair, but it only prevailed for a few days, when through certain underground workings unknown to these officials and through certain secret and mysterious influences this policy was changed. The commanding officer then announced that *any non-union miner desiring to enter the employment of the company would be protected by the national guard.* In other words, the operators could recruit unfair labor in any quarter. They could ship in the strike breakers from surrounding states and receive guard protection in so doing. . . . Then the United Mine workers *demanded the right to conduct peaceful picketing under the conditions defined by the highest courts of the land.* But this right was denied them. They then asked that they be given the right of free speech, that they be given the right to hold public meetings and to advocate the cause of labor, *that they be given guard protection in the exercise of these rights*, but *they were denied these rights.* In other words, the guard became, as the operators desired it should become, *a weapon in their hands to destroy the morale of the striking miners*; that *it be used to protect the strike breakers and to safeguard the operators in the enforcement of their policies designed to destroy organized labor.*[59]

The National Guard troops were withdrawn from Carbon County :tween August 13 and September 15, 1922. The expenses for maintaining the Guard in Carbon County for the three-month period amounted to nearly $75,000.[60]

After a long and bitterly contested struggle, the Utah strike was quickly ended upon settlement of the nationwide bituminous coal strike. On August 7 John L. Lewis signed with operators responsible for producing 60 percent of the nation's bituminous coal. Other operators soon followed, and on August 16 Lewis ordered his union men back to work. The UMWA succeeded in preventing the proposed wage reduction that had precipitated the strike. The union and the coal companies agreed to accept the 1920 wage scale until April 1923. On September 1, 1922, the Utah coal operators announced that a wage increase effective immediately would be granted voluntarily to their men that would "approximate the average wage scale in effect in [the] eastern and western field." The raise amounted to nothing more than a restoration of the wage scale in effect before the 30 percent reduction in April. In addition, the operators promised to reward the loyalty of those men who had worked during the five-month strike with a bonus.[61]

Union officials previously had requested to meet with operators on September 2 to discuss a strike settlement; but the operators' announcement regarding the new wage scale and bonuses, coming the day before the proposed meeting and emphasizing that the wage increases were being made "voluntarily," left little doubt that the operators' strategy remained one of avoiding any form of direct or indirect recognition of the UMWA.

The new wage scale provided a minimum wage of $7.00 a day for outside men and $7.95 a day for inside men. The scale, Utah operators declared, was "the highest ever paid in Carbon County."[62]

To justify the wage adjustments, the price of coal to the consumer increased from $9.50 to $10.00 a ton. State officials objected to the increase, claiming that the operators had pledged no price increase would take place if the National Guard were sent to Carbon County. The public questioned why, with the same wage scale and freight rates in effect as before, the fifty-cent increase was justified. Frank N. Cameron, vice-president and general manager of the Utah Fuel Company, acting as spokesman for the operators, declared that the rate increase was necessitated by overhead charges, taxation, a shortage of coal cars, and "other matters." The newspapers countered by asking why Wyoming coal, produced under schedules as high or higher than those in Utah and costing twenty-five cents a ton more to be shipped to Salt Lake City than coal from Carbon County, sold for a dollar less per ton than Utah coal. Governor

Mabey, in calling for an investigation of the price hike, calculated that if the Utah mines produced approximately 500,000 tons of coal, as they had in the preceding month of August, then the consumers would pay the operators $250,000 more per month ($3,000,000 more a year) than the coal companies were entitled to receive. Mabey also suggested that, since the price increase had been announced simultaneously by all operators, there might be a matter of illegal combination in restraint of trade.[63]

A grand jury impaneled to investigate the coal-price increase agreed with Governor Mabey and brought indictments against F. N. Cameron, general manager of the Utah Fuel Company; F. A. Sweet, president and general manager of the Standard Coal Company; Moroni Heiner, vice-president and general manager of the United States Fuel Company; F. H. Rolapp, president of the Royal Fuel Company; J. W. Knight, general manager of the Knight Fuel Company; and J. H. Tonkin, general manager of the Independent Coal and Coke Company.[64]

Although union officials realized that the goal of union recognition had not been reached, they encouraged their men to return to work following the clarification of the new wage scale. The miners were, however, urged to maintain their affiliation with the UMWA and to encourage others to join the organization in anticipation of future confrontations with the Utah coal operators.[65] When the strike ended, the UMWA did not abandon the Utah fields as they had following the 1903–4 strike. Instead, they remained to provide legal and monetary support for those who had been arrested during the strike.

The first trial involved George Manousos who was charged with the attempted murder of Sam Dorrity. Despite the claims of eighteen defense witnesses that Dorrity fired the first shots, Manousos was found guilty and sentenced to twenty years in prison. The next five trials involved eight Greeks charged with the murder of Arthur P. Webb. Although the state was willing to try all eight together, attorney Samuel King insisted on separate trials. King reasoned that the public would soon grow weary of the trials, and in the end the men would either be given reduced sentences or be acquitted. The defense strategy worked. In the first two trials, Price juries found Pete Kukis and Mike Zulakis guilty of murder and gave each a life sentence. By the third trial, however, the outlook began to brighten. Attorney King obtained change of venue to Castle Dale for Mike Pagialakas. Although he, too, was found guilty, his sentence was reduced to ten years in prison. Still, King was not satisfied with Emery

County as a trial site. Claiming that the influence of the coal companies was as strong there as in neighboring Carbon County, King obtained another change of venue; and the fourth trial, that of Tony Kambourakis and John Kriaris, was held in Salt Lake City. The two men were found guilty of voluntary manslaughter but received the minimum sentence of one year. The fifth and last trial was also held in Salt Lake City where John Dantis, George Spetris, and Steven Lakakis were acquitted. Following the acquittal, bail for the remaining eight strikers charged with the murder of Webb was reduced; and in November 1925, charges against all eight were dropped. On March 20, 1926, paroles were granted to those in prison.[66]

Although UMWA efforts to secure justice for the sixteen men had taken nearly four years and had cost a great deal of money, the UMWA commitment proved worth the effort. To the union, the trials were a conspiracy by the coal operators; not to punish the men for murder, though perhaps some were guilty, but rather to weaken the influence of labor unions. The expenses of the strike and trial nearly bankrupted the District 22 treasury. A total of $73,881.81 was sent to Utah between April and December of 1922. In addition, nearly $70,000 was spent for legal service and expenses through November 1, 1923. Assessments were levied for eight months against the District 22 members, nearly all of whom were employed in the Wyoming coal fields. However, the funds raised fell short of the amount needed to pay the Utah expenses, and district officials were forced to seek assistance from the International Executive Board of the United Mine Workers of America. The board immediately provided $20,000 to assist with the Utah expenses.[67]

In the brief submitted to the United States Coal Commission, attorneys representing the Utah coal operators concluded that the Utah strike had been forced on the state in an effort by the UMWA to keep nonunion coal out of the market during the nationwide strike.

> Accordingly, its paid agitators invaded a peaceful community, stirred up strife, fomented violence and discord, endeavored to strike terror into the hearts of a susceptible foreign population, used their utmost efforts to prevent the interference of the state authorities to maintain law and order, and finally created such a condition of lawlessness and disorder that the state authorities had to step in with their troops to bring about the peace and quiet and orderly government which had existed before the invasion.[68]

UMWA tactics in Utah, according to company attorneys, indicated that the union's policy and methods were un-American, undemocratic, and "a distinct menace to our American form of government and its institutions."[69]

From today's perspective, the Carbon County miners, faced in 1922 with a 30 percent wage reduction, seemed justified in going out on strike. Although UMWA officials from District 22 were in Utah at the beginning of the nationwide strike on April 1, 1922, seeking to strengthen the union organization among the Carbon County miners, those officials believed that union strength in Utah was still too weak to include the state in the nationwide strike call. However, the Utah miners, responding to the wage cuts and to promises of better protection against the coal company abuses through strong union organization, struck without official sanction. The Utah strike was, therefore, not the result of outside agitators who stirred up the previously contented miners as the coal operators claimed but rather a response to legitimate economic concerns.

Violence and intimidation were not policies advocated by UMWA organizers. Through long experience they had learned that violence would quickly destroy public sympathy for a strike. They continuously cautioned their men to avoid violence and intimidation. The shooting incidents that occurred can be traced as much to coal company guards and officials as to strikers.

Although the miners gained a restoration of the pre–April 1 pay scale, the strike failed on several counts. The restored pay scale was short-lived, and the miners were forced to accept pay reductions in 1925, 1928, and 1931. Perhaps more important than pay reductions was the issue of union recognition. Union officials realized its importance at the beginning of the strike when they offered to accept the wage reductions if the union would be allowed to carry out organizing work among the miners. Although District 22 had accepted jurisdiction of Utah in 1921 with high hopes for success in establishing a strong union in the previously unorganized Utah fields, it merely maintained a caretaker organization after the strike. To the union's credit, however, it did not abandon the miners as it had following the 1903–4 strike. District and national officials and the union miners in Wyoming accepted the financial obligations from the defeated region without complaint.

Governor Mabey deserves praise for refusing to be stampeded into calling out the Utah National Guard. However, the question remains unanswered whether his position was dictated by his program of reduced state spending or by sympathy with the striking miners. Mabey displayed the same nativistic sentiments that his predecessor, Heber M. Wells, had shown in 1903. Speaking to the legislature in January 1923, Mabey explained why he called out the National Guard.

> An intolerable condition had arisen, not entirely due to the strike, but largely because of the many people of foreign birth and traditions who had not yet learned proper reverence for American laws and American ideals. These had flaunted their disrespect for our established order in the face of authority and engendered a feeling of terror among the more timid members of society.[70]

Nevertheless, the bitter attacks by the local papers against Mabey help to explain his poor showing in Carbon and Emery counties during his unsuccessful bid for reelection in 1924.[71]

The National Guard initially showed great potential as a neutral law enforcement body; however, its policies soon supported the coal operators' efforts to bring in strikebreakers. The National Guard policy was one of the most important reasons why the union did not push for other concessions once the wage scale had been restored.

As in earlier strikes, the foreign-born miners proved to be the backbone and heart of the union organization. This, however, also proved a liability as a few of the immigrants overstepped the bounds of proper conduct and gave some misinterpreted credence to the charges of radicalism and anarchy. Charges against the foreign-born miners surfaced in an era highlighted by nativism, the Red Scare, and a general spirit of intolerance raging throughout the country. In any case, the stigma of "foreigner" or "striker" did not ameliorate or explain the potential hazards of coal mining.

PLATE I: *Coal miners in the Scofield area about 1900. Note the young boys standing on coal car. Courtesy of Utah State Historical Society, Salt Lake City, Utah.*

PLATE III: *Company houses at Sunnyside about 1900. Courtesy of Utah State Historical Society, Salt Lake City.*

PLATE III: *Company houses at Sunnyside about 1900. Courtesy of Utah State Historical Society, Salt Lake City.*

PLATE IV: *Winter Quarters explosion, May 1900. Removing bodies from the mines. Courtesy Utah State Historical Society, Salt Lake City.*

PLATE V: *The Luoma residence in Winter Quarters, May 1900. Mr. and Mrs. Abe Luoma, seated in the foreground, were recent immigrants from Finland and lost six sons and three grandsons in the May 1, 1900, Winter Quarters explosion. Courtesy of Utah State Historical Society, Salt Lake City.*

PLATE VI: *Home with three burial caskets in the front yard, Winter Quarters explosion, May 1, 1900. Courtesy of Utah State Historical Society, Salt Lake City.*

PLATE VII: *Provo burial detachment which volunteered to dig graves in the Scofield cemetery for victims of the Winter Quarters explosion, May 1, 1900. Courtesy of Utah State Historical Society, Salt Lake City.*

PLATE VIII: *The Bullpen in Price, Utah, in which 120 Italian strikers were incarcerated during April 1904 during the 1903–4 coal miners' strike. Note the rifle-carrying guards. Courtesy of Utah State Historical Society, Salt Lake City.*

PLATE IX: *Nephi Company of the Utah National Guard in front of the Clear Creek Finn Hall during the 1903 coal miners' strike. Courtesy of Utah State Historical Society, Salt Lake City.*

Plate X: *Man Trip probably at Hiawatha. Courtesy of Utah State Historical Society, Salt Lake City.*

PLATE XI: *Convention of the United Mine Workers of America held at Helper, Utah, June 8, 1919, during the heyday of union activity in Utah during and just after World War I. Courtesy of Special Collections, Marriott Library, University of Utah.*

PLATE XII: *Company houses at the coal town of Consumers, March 1936. Photograph courtesy of Dorothea Lange.*

PLATE XIII: *Trucks loaded with coffins of Greek miners killed in the March 1924 Castle Gate mine explosion en route to burial in the Price cemetery. Courtesy of Utah State Historical Society, Salt Lake City.*

PLATE XIV: *Castle Gate March 1924 after news of the mine explosion in the Castle Gate No. 2 mine which claimed 174 lives. Courtesy of Utah State Historical Society, Salt Lake City.*

Chapter VIII

The Castle Gate Mine Disaster

The men who answered the work call in Castle Gate on Saturday, March 8, 1924, considered themselves fortunate. Theirs was only the third eight-hour shift to work during the month. The miners were, for the most part, married men and older employees who had been given work over younger, single men in the No. 2 mine when the No. 1 mine closed March 1 because of reduced coal orders.

At 7:30 a.m., within an hour of entering the Castle Gate No. 2 mine, the entire work force of 171 men was dead; the men were victims of two violent explosions that shot through the mine turning the main tunnels into gigantic cannon barrels.

Tests by the United States Bureau of Mines showed that Castle Gate coal produced the most flammable and explosive coal dust in the country. The Castle Gate mines had pioneered the blasting practice of shooting down coal when workers were outside the mine, and following the tragic explosion at Winter Quarters in 1900, the eastern Utah mines had given careful attention to developing sprinkling systems to wet down and check the volatile coal dust. However, a combination of human mistakes and natural factors produced the fatal circumstances.

The explosion was caused by an open flame that ignited methane gas, which in turn ignited the excessive coal dust, producing a chain reaction of exploding coal dust. Reports concluded that there were areas of the mine that had not been properly sprinkled and that coal, spilled from overloaded cars moving through the mine, was pulverized into dangerous dust by passing men, animals, and cars. The first hour of work was usually hazardous; the miners first shoveled the coal that had been shot into available cars and then turned to sprinkle their section of the mine until more cars arrived for loading. It had been several days since the mine had been worked, accounting for a greater accumulation of coal dust; and although sprinkling was important in controlling dust, it was not 100 percent effective.

Although Utah coal operators were proud of their record since the 1900 explosion in preventing similar occurrences, primarily through the practice of sprinkling, it was later acknowledged that this practice could not cover all areas of the mine. In some locations that were sprinkled at the beginning of the day, coal dust levels reached dangerous proportions by the end of the workday. Rock dusting, it was belatedly concluded, was a much superior means of controlling coal dust. The rock dust could be applied to all areas and was not affected by the dry Utah climate or by frozen or broken pipes that constantly interrupted sprinkling operations. The rock dust formed a more permanent protection lasting weeks and even months.

The first explosion was ignited in the No. 2 room. Top coal had been shot down the night before, and the fire boss had found gas near the roof. When he climbed on the pile of coal to determine how much gas hovered near the ceiling of the room, his lamp extinguished. The fire boss climbed down and was in the process of relighting the lamp when he or another miner in the area ignited the gas. Either by the actions of the fire boss or as a result of a fall of coal that occurred just before the explosion, the gas traveled down from the ceiling and contacted the open flame. The initial explosion sped down room No. 2 to the entry where it split, the main force traveling through the mine and eventually out the mine entrance. The secondary force swept deeper into the mine; but by the time it reached the 5th left raise entry it had dissipated and only extinguished the lights of men working that section of the mine. However, when these men struck matches to relight their lamps, they ignited the coal dust, and a second explosion roared back along the course of the first. Within a few minutes, all 171 men were dead. They were killed by the force of the explosion rather than the usual carbon monoxide, or afterdamp, which was generally the most deadly factor in mine explosions. However, carbon monoxide was present, and as T. L. Burridge, a witness to the explosion, recorded in his diary, "Smoke and gas was coming from all three entrys so thick that if a man walked up to the mouth of any of these entrys and stood there for two or three minutes he would have been overcome with gas."[1]

Ironically, the second explosion might not have occurred if the miners had been wearing the new electric-powered headlights purchased by the

Utah Fuel Company but not yet issued because the charging apparatus had not arrived.

The force of the explosion blew out timbers and caved in the main portal for a distance of about thirty feet. Eight-inch water pipes were "turned out like straws"; emergency gates made of steel rods six inches in diameter "were twisted and thrown to one side as though they had been made of wire." The insulation on electric wires was blown off by the blast; timbers and rails were thrown an estimated five thousand feet across the canyon; and coal along the walls of the tunnel was coked from the heat of the explosion.[2]

The 171 dead miners represented 12 percent of the total population of Castle Gate. One hundred and fourteen of the men were married and left a total of 415 dependents. The miners' nationalities reflected the diversity that characterized the Utah coal fields and, according to contemporary reports, included forty-five Americans, two of whom were black, fifty Greeks, twenty-five Italians, thirty-two English and Scots, twelve Welshmen, four Japanese, and three Austrian or present-day Yugoslavians.[3]

The dead included such men as Benjamin Thomas, the Mormon bishop of Castle Gate and fifty-five-year-old native of Wales, who came to Castle Gate by way of Wales, Utah; Andrew Gilbert, a seventy-three-year-old veteran of Utah mining, who had survived the Winter Quarters explosion a quarter of a century before; and a black, Prince Alexander, who arrived in Castle Gate in 1914, stayed to work in the mines, and earned the deep respect and admiration of the coal camp for his valiant and unselfish efforts during the 1919 flu epidemic.[4]

Shortly after the explosion, rescue workers were rushed to Castle Gate from the surrounding mines. They faced a difficult task: debris and fallen rock clogged the entry and passageways, and deadly carbon monoxide required all rescuers to wear cumbersome oxygen breathing apparatus. As rescue work began, one of the workers discovered he had a faulty breathing apparatus. George Wilson, captain of the Standardville team, attempted to bring the man out of the mine. Struggling for breath, the miner accidentally knocked Wilson's nose clip off causing him to inhale deadly gas. The rescue team took the stricken man from the mine and sent in another team for Wilson. Wilson had begun his mining career at the age of fourteen in the Clear Creek mine and was no stranger to disaster. He had lost three brothers and had seen his father and another brother suffer severe injuries in the Winter Quarters disaster.[5] Although the rescue workers reached him and had him outside the mine in fresh air within five minutes, George Wilson died bringing the total dead to 172.

The rescue teams came from Storrs, Hiawatha, Clear Creek, Kenilworth, Wattis, Mutual, Columbia, Winter Quarters, Rolapp, Scofield, Standardville, Mohrland, Peerless, Sunnyside, and Castle Gate.[6] In addition, the Union Pacific coal mine at Rock Springs, Wyoming, offered to fly its rescue team to Castle Gate, but its assistance was deemed unnecessary.

The rescue teams were divided into three shifts of eight crews equipped with oxgyen-apparatus systems. In each shift, the crews were split with half working and the other half on standby at the fresh-air base. Each rescue crew was pressed into service within two hours after the first crew's oxygen supply was depleted. Forty-three canaries were donated by Salt Lake City residents and pet store owners to assist the rescue workers detect the presence of the deadly gas. The system was simple: "Each exploring rescuer advanced along the tunnel with a bird cage in his hand. When his feathered companion collapses he knows he is in a danger zone." Newspaper readers were reassured that unless the gas was suddenly encountered and extremely dense, "the bird will recover under treatment."[7]

Initially, the rescue workers were spurred on by the hope that some of the miners had survived the explosion and had been able to successfully barricade themselves from the deadly gases. Workers and relatives assured themselves that the 171 men were experienced miners with knowledge of the mine and who knew how to take care of themselves. A pile of work tools and another pile of lunch boxes were found, suggesting that the miners had placed them together before seeking safety.[8] The rescuers did not understand the nature of the two explosions and presumed that portions of the mine had been spared the violent explosion evident at the mine entrance.

Hope soon turned to a realization that none of the 171 men would emerge alive from the shattered hole, and as rescue changed to recovery, several obstacles were encountered. Carbon monoxide made the use of oxygen apparatus mandatory; numerous fires within the mine had to be extinguished; debris and fallen rock had to be cleared; and in low places of the mine, ponds of water had formed when sprinkling pipes and pumps used to remove the natural accumulation of water were destroyed. Water depths of five to ten feet were reported in some places.[9]

As the recovery work progressed and as bodies were brought from the mine and taken to the makeshift morgue in the Amusement Hall, the identification of the dead proved difficult. In the company office building, the checkboard usually used to identify who was in the mine had been knocked from the wall by the explosion, and company officials were obligated to conduct a thorough canvass to prepare the list of men who entered the mine that morning of March 8. The bodies that were recovered were badly mutilated and, in some cases, dismembered. David Jones, who had been transferred as chief clerk at Castle Gate to another Utah Fuel Mine in Del Carbon, Colorado, was sent for by telegram and was asked to return and assist in the identification. Alma Hardee was identified when his overalls, shoes, and socks were sent to his house for his wife to examine. A day after the unseen corpse of Alma Hardee was buried, his head was recovered from the haulage way; the body was exhumed, and the skull placed inside the casket.[10] In another unpleasant episode, a near altercation took place between Greeks and Japanese when a body, originally identified as that of Pete Dunis, was returned to the morgue while en route to the cemetery and was reidentified as Tom Tageuchi.[11]

The relief work performed by friends, volunteers, and the American Red Cross matched the miners' heroic rescue and recovery efforts. A reporter for the *Salt Lake Telegram,* writing from Castle Gate, noted: "There are many here who will bless the name of woman. . . . As angels of mercy they have gone from home to home and have attempted to ease the burdens of grief."[12] These women cooked meals for both the rescue workers and families of the dead, tended children, fed babies, performed household chores, sewed and purchased black mourning clothes, and assisted in every way possible to alleviate and soothe the suffering and grief that blanketed the town. Many of the women worked under the direction of the Carbon County chapter of the Red Cross that directed the relief effort until a committee was appointed by Governor Charles Mabey to report on the situation of the widows and orphans at Castle Gate.[13] The governor's committee, chaired by Imer Pett, general manager of the Bingham Mines Company, also included Carl R. Marcusen, a Price banker and merchant, Mrs. C. H. Stevenson, chairman of the Carbon County chapter of the American Red Cross, Frank Hanson, claim agent of the Utah Light and Traction Company, and Darrell T. Lane, Utah commander of the American Legion.[14]

Others visited Castle Gate to assess the situation firsthand, including Governor Charles Mabey; Mormon church President Heber J. Grant; Archbishop Peter Fumason-Blondi, apostolic delegate from Washington; Bishop Joseph S. Glass of Salt Lake City; Italian Consul, Fortunato

Anselmo; Utah Fuel Company officers; and officials from the United States Bureau of Mines.

Because of the generally mutilated condition of the corpses and the fact that recovery took nine days, the dead were buried as quickly as possible. Caskets were sent from Salt Lake City, and 125 graves were dug in the Castle Gate cemetery—a generous number considering most of the men were buried in their hometowns or other communities. Services for the fifty Greeks were held in the Assumption Hellenic Orthodox Church in Price, and most were buried in the Price cemetery. Other burials were held in Kanarraville near Cedar City, Wales in Sanpete County, Orangeville in Emery County, Salt Lake City, Spanish Fork, Price, Helper, Nephi, Mt. Pleasant, Heber City, and Fort Smith, Arkansas. Castle Gate services for the dead were conducted by the Reverend R. C. Jones from the Price Methodist Church; two Episcopal ministers, Reverend W. F. Bulkley from Provo and the Reverend C. H. Johnson from Price; Mormon Bishop George Ruff; and Catholic Archbishop Peter Fumason-Blondi and Bishop Joseph S. Glass.

As the recovery and burial efforts continued, primary concern was given to the widows and dependents left by the dead miners. The widows included both those who had been married only a few weeks and those who were veterans of the Utah coal experience like Mrs. George Shurtliff who lost her first husband, David Evans, in the Winter Quarters explosion of 1900.[15] Two women claimed to be the widow of Ed Willis. The first said she had married Willis in 1917 but had not lived with him for some time; the second said she had lived with Willis for the past several years. The women, it was reported, quickly became good friends and walked arm in arm to Willis's burial in the Castle Gate cemetery. However both asserted they were entitled to compensation for Willis's death.[16]

Somber notes of grief were sounded for the married Greek, Austrian, Serbian, and Italian miners. Their wives spoke little English and were still adapting to the new customs of America and harsh coal-camp life. When Fortunato Anselmo, Italian consul, arrived in Castle Gate, he found a family of Italian children who had not been fed for several days. Another Italian widow lamented the fate of her husband by chanting, "O dola mia, O dola mia" ("O my sorrow, O my sorrow") over and over.[17] Others also vented their grief in their native tongue, while relief workers, sensitive to the importance of black mourning attire for the foreign-born widows, made gallant efforts to provide the symbolic apparel.

The number of dependents left by the dead totaled 415, including 114 wives, 239 children between the ages of one and eighteen, 25 unborn children, and 37 other dependents such as parents, brothers and sisters, older children, and other relatives. Of the 239 children, 180 were ten or younger.[18]

Not understanding the scope and full meaning of the disaster, the children were especially affected. One five-year-old-girl, Maria Tagliabue, concerned that her mother had not received word of her father, went to the Amusement Hall and persuaded the workers there to allow her to search the personal items and clothing of the dead for some identification of her father. As she examined the piles of items:

> Suddenly Maria's large black eyes were strained and opened wide and with a frightened expression she turned to her escort "These are my daddy's socks . . . I know that the mending was done by mama." Then realizing that she had really found her daddy little Maria's eyes filled with tears and muffled sobs racked her until one of the Star Legion mothers gathered Maria in her arms and carried her from the scene.[19]

Saline Hardee Fraser, thirteen years old at the time she lost her father and grandfather in the Castle Gate explosion, recalled many years later how she and another girl occupied the long hours of waiting by bouncing a ball. Suddenly someone

> came and snatched the ball away from us and bawled me out and said I should be ashamed of myself; but I remember that ball and the girl brought it; we were just trying to occupy ourselves. At the time I didn't think I was really doin' anything bad. There was nothing we could do. We were just waitin' and I felt as bad as anybody. . . . I didn't see what I was doin' wrong, but I remember gettin' bawled out.[20]

The relief committee investigated the financial circumstances of the dead miners' wives and adult dependents and classified ninety-four as poor, nine as fair, and the remaining unknown. One hundred-forty-three dependents were without income and did not own their residences. One dependent owned a small house and one widow was a schoolteacher. The condition of the others was unknown.[21]

The report illustrated the precarious financial circumstances under which most Utah coal miners lived. The miners did not own their homes, usually had no other income than what they earned mining, often had substantial debts at the company store, had little (if any) savings, and most carried no life insurance.

Given the bleak financial status of the dead miner's families, the provisions of the State Workmen's Compensation Law were a godsend. Death benefits provided families with 50 percent of the worker's weekly wage up to $16 a week and $5,000 for a period of six years. In addition, the law provided for funeral expenses up to $150. Since the Utah Fuel Company had opted to carry its own workmen's insurance, it paid the compensation benefits; the total cost to the company approached $750,000.[22]

In addition to the workmen's compensation benefits, approximately $130,000 was raised by private donations throughout the state and surrounding areas. Nine trustees appointed by the governor administered the relief fund.

While the commission, known as the Castle Gate Relief Committee, administered the relief fund on a voluntary basis, the actual relief work was carried out by Mrs. Annie D. Palmer, a welfare worker whose salary was paid by the relief fund. Mrs. Palmer worked directly with the dependents and widows to assess their needs, to provide counseling, and to assist in the application for emergency funds, such as educational or medical expenses, which had to be approved by the committee. She also maintained case files on each of the widows with whom she worked. The files contain details of the problems encountered and the adjustments made as the women and their families sought to rebuild their lives after the March 8 tragedy. The reports tell of immigrant women who intended to return to the old country, assuming they would be paid money they could use as they wanted rather than on railroad or steamship tickets. Instead, Mrs. Palmer made arrangements and purchased tickets herself, encountering considerable difficulty when apparent second thoughts about returning to Europe surfaced.[23]

Some women turned to alcohol for solace, and when it was learned that money for food and clothing was being spent for alcohol, orders for groceries instead of cash were sent.[24] For many who carried large debts accumulated before the disaster, the situation was bleak. Several women tried to sell raffle tickets for cars that they possessed.[25] One widow saw the explosion and ensuing relief as a boon since her husband was old and soon to die. Had he died at home in bed, which was likely to be very soon, they would have been left without a dollar for support.[26]

Most of the entries in Mrs. Palmer's case files are concerned with health problems and family difficulties. Representative of these entries is the following:

> Mrs. R. had had teeth extracted, but still has much pain in gums. Dr. Evans taking care of her. She was very much worried about family cares. Has not heard from Iris for six weeks. Orson staying out late at night, disobedient, and untruthful; Joe following in the way of Orson. Eliza and Viola do not come home when school is out and tell each other to hide from her. She thinks she will let Joe go to live with a relative on a farm, and perhaps one of the little girls will go to a brother of her husband in Salt Lake. She stated she would rather see her boys laid by their father than to have them grow up to dishonor her. She is sure Iris gambles and probably drinks. The only one in whom she has hope is LaFayette. He does all he can for her, and realizes her need. But he will not tolerate the actions of the younger boys without trying to whip them into line. She was advised that if Orson does not do better she may be under necessity of sending him to Industrial School.[27]

In time, adjustments were made. By 1936, when the relief fund payments ended, sixty-one of the widows had remarried and eleven had died. A total of $132,145.27 had been disbursed from the fund by 1935 with the amount broken down as follows:[28]

Monthly Relief Payments	$91,211.65
Emergency Relief	$17,488.25
Administration Expenses	$12,445.13
Final Settlements to Those Leaving To Return Home	$11,000.24

With the removal of the last bodies from the Castle Gate No. 2 mine, the Utah Fuel Company began rehabilitation work to reopen the mine. The cost was estimated at $250,000, bringing the cost of rehabilitation expenses and workmen's compensation benefits to approximately one million dollars.[29] Until rehabilitation was completed, the old Castle Gate No. 1 mine was reopened. There was no dearth of workers for the Castle Gate mines. A substantial work force still remained in Castle Gate, and a large influx of unemployed miners from surrounding camps streamed to Castle Gate even as the recovery of the victims' bodies continued.[30]

Coal operators and government mining officials, shaken by the magnitude of the disaster and the failure of what were long-accepted safety procedures, met in Salt Lake City on April 2 to amend the coal mine safety orders issued by the State Industrial Commission. The amendments and resolutions called for the legislature to fund more coal mine inspectors; to limit the size of explosives used in blasting coal; to require the firing of shots by electricity when all men, except the shot firers, were out of the mine; to stipulate that men be equipped with electric lights and that flame lamps be forbidden, except for testing purposes; to make rock dusting mandatory on all main entries and slopes with rock dust barriers preventing any explosion from extending to other parts of the mine; to require that water be supplied with all cutting machinery and loading machines to keep down the coal dust; to provide for adequate sprinkling to wet down the coal dust at all working faces; to prohibit the moving of methane gas accumulations while a shift was in the mine; and to require that all coal dust be cleaned up and loaded out from all abandoned rooms, air courses, and other places.[31]

Although the new safety precautions did not prevent further mine disasters, they helped minimize their seriousness. On February 8, 1930, twenty-three men were killed at the Standardville mine when a spark from a cutting machine in the No. 3 slope caused an explosion killing three men outright and spreading the deadly carbon monoxide gas through much of the mine. Seventeen miners were killed by the gas, and later three rescue workers, by falling rock. The rock dusting prevented the Standardville mine from becoming another Castle Gate "cannon" because the explosion was confined to one location in the mine. Nine men were saved, two by rushing from the mine when the explosion occurred. After twelve hours, five miners in the No. 1 slope were rescued by search parties carrying the traditional canaries to avoid the dangerous carbon monoxide. The five men had barricaded themselves with canvas and maintained a continual conversation to monitor the spread of carbon monoxide. When one of the men could no longer talk, they knew it was time to leave a message chalked on the wall and move to another room.

Once again the seventeen dependents of the twenty-three dead were provided with sixteen dollars a week for 312 weeks under provisions of the State Workmen's Compensation Law. Among those covered were the wife and four children of Wataru Ishimue, who had written on the leather visor of his miner's hat four characters that, when translated, read, "Dear brother, please support my wife and children."[32]

The aftermath of the Castle Gate disaster differed in many ways from the Winter Quarters disaster. No vocal labor protest grew from the misery and fear that followed. The abundance of experienced unemployed miners provided an instant labor force, unlike 1900 when new workers had to be found for the mines. Responsibility for the disaster was clearer with both the miners and coal company responsible for the combination of factors that resulted in the grievous tragedy. The relief program was much better organized and much more effective at Castle Gate than at Winter Quarters. Castle Gate was more centrally located, in close proximity to Price and Helper. In addition, the organized efforts of the Carbon County chapter of the American Red Cross and Governor Charles Mabey's involvement and leadership in appointing an effective relief committee all helped ameliorate the tragedy. The provisions of the compensation law were immediate; and, for the most part, compensation and relief payments were carried out with commitment, judiciousness, and wisdom. Unlike 1900, where the death benefits were paid in one lump sum, the Castle Gate Relief Commission was careful to insure financial assistance for those in serious need until 1936. In addition to sending monthly cash payments, the Castle Gate Relief Commission exercised prudent judgment in providing money for medical, dental, educational, and other legitimate expenses. Thus, support continued for the widows, orphans, and dependents during the bleak 1920s and the difficult years of the Great Depression. Although the Scofield disaster was commemorated each May 1 with services at the Scofield cemetery, the situation was much different in Castle Gate. After marking the first anniversary of the tragedy, it was decided to discontinue future ceremonies since they "could never be anything but painful to . . . [the] participants."[33] The Castle Gate disaster left twenty-eight fewer dead than the Winter Quarters calamity with its 200 dead. With the deaths, six years later in 1930, of the twenty-three miners at the Standardville mine and with the turmoil and upheaval of 1933, the Castle Gate catastrophe was obliged, in the perspective of many, to share the limelight with other tragedies and upheavals in the decade that followed.

Chapter IX

Retrenchment and Disenchantment, 1924-32

Throughout the remainder of the 1920s, union efforts in Utah coal fields were subject to much the same intolerance and reaction that characterized the rest of the country. In the first legislative session to be held after the 1922 strike, a Right-to-Work Act was passed and signed by Governor Charles Mabey that guaranteed workers the right to employment without union affiliation even though a majority of the shop or industry workers had become organized. Moreover, union leaders and members were prohibited from almost any attempt to encourage their fellow workers to join the union.

Persons violating the law were guilty of a misdemeanor and subject to a fine not to exceed $300 or to imprisonment for a period not to exceed six months or to both. The *Wyoming Labor Journal* declared that the Utah lawmakers had passed a law to "protect scabs" and that the state was governed by an unholy combination of big business and the Mormon church, which the *Journal* considered nothing more than "a highly commercialized church organization which teaches its converts that it is not Godly to demand more than the employer is willing to give."[1]

Although the open-shop movement was the primary weapon against unionism, Utah coal operators considered other methods for eliminating the unions from the coal industry. In an address to the American Institute of Mining and Metallurgical Society in October 1924, Thomas A. Stroup, superintendent of the Utah Fuel Company mine at Clear Creek, argued that through a complete mechanization of the mines this could be accomplished. Stroup declared that labor relations in the coal industry could only function if such "artificial handlings of the check off of union dues and certification of miners by the state . . . [were] swept away."

Obviously Stroup's scheme would leave the coal miner completely at the mercy of the employer. However, according to Stroup, this would be the most advantageous situation for the miners because they were part of the 80 to 90 percent of the population who were incapable of "clear thinking or self direction" and made up the "preponderance of the subnormal in our working population."[2] Categorized in these demeaning terms, the average worker was expected to be like clay in the hands of the skillful union organizer.

Defeated in the crusade for union recognition during the 1922 strike, Utah coal miners still hoped during the remainder of the decade to establish a union eventually. District 22 officers from Wyoming made periodic visits to Utah. In Helper a local union of the UMWA, No. 1984, survived and included members coming from the surrounding company towns where locals were not tolerated. Frank Bonacci, employed at the Spring Canyon mine, served as financial secretary for the local union and collected dues from several Utah miners.[3] Another local union was also maintained in Scofield. The two communities of Scofield and Helper continued as sanctuaries for the miners from the ever-watchful and often-repressive coal company.

In April 1924, the *Wyoming Labor Journal* carried an article discussing the Pahvant Coal Company's proposal to sell bonds to "union and liberal minded people of Utah" and then to operate its mines in Spring Canyon under an agreement with the UMWA.[4] Although the plan had the endorsement of at least one official of the Utah Federation of Labor and although many active Salt Lake and Ogden union men were reported to have invested in the company, the scheme failed.

According to labor spies, the UMWA was not the only group to show an interest in the Carbon County miners. In a letter to Thomas R. Stockett, general manager of the Spring Canyon Coal Company, dated August 12, 1924, Arthur McArthur, district manager of the Globe Inspection Company, reported that the Industrial Workers of the World was preparing for an active campaign among the Utah coal miners during the coming fall and winter with the hope that by the spring of 1925 it would be able "to dictate terms of employment." The letter quoted a report submitted August 2, 1924, by the Globe Company's inspector.

> While in Price a few days ago I met two men who were not at all afraid to acknowledge membership in the I.W.W. organization. At the time they were working in Helper but said they intended to get on at one of the coal mines just as soon as work picks up. They referred to the United

> Mine Workers of America as a "big joke" and predicted that within the next year the coal miners of this state would be lined up solidly under the I.W.W. banner. Judging from what they had to say, this organization has already been successful in locating a number of delegates throughout the Utah coal fields and who are working under cover. They didn't mention any names or any particular camps; in fact, were rather backward about giving out any information. I was unable to get the names of these fellows at the time but will be on the lookout for them in case they attempt to locate in my camp.[5]

Three weeks later, on August 21, an inspector assigned to the Sunnyside area, reported on IWW activity.

> Seemingly the I.W.W. organization is plugging away in its effort to locate as many members as possible in the Utah coal camps. A few days ago I came in contact with a very radical member of this organization but who has since left camp, headed for Columbia. He rode into camp on a freight car and the next morning when he went up town was met by what he thought to be two of the camp guards. When I met him later in the day he asked, "What kind of a place is this?" When I asked him for an explanation, he replied, "This morning I was stopped by a couple of fellows who wanted to know where I came from, where I was going, what I wanted, etc., in fact, they wanted a history of my past life and suggested that I had better look elsewhere for work. Is this camp full of bulls and spotters?" He then said, "Sometime in the near future some of these boobs will wake up and carry a good red card." I told him he was in the wrong place to start anything like that as a majority of the men in this camp appear to be quite well satisfied and didn't care to be mixed up in any radical move which would only make conditions worse than they are. He didn't like this line of talk, said he was leaving camp as quickly as possible and intended to rustle Columbia.[6]

No other evidence links the IWW to the Utah coal fields during the 1920s, and perhaps the situation was misrepresented or overstated by the Globe Inspection Company representatives, who were attempting to convince General Manager Stockett to hire an inspector for his mines. Nevertheless, the labor spies of the Globe Inspection Company were found in the major coal camps of Carbon County, and at least one, the agent at Castle Gate, was a close acquaintance of Frank Bonacci and partook of the Italian organizer's misplaced confidence regarding union matters.

In 1925, Tony Radalj, international executive board member from UMWA District 22, renewed the contacts he had made in Utah during the 1922 strike and for the next several years was the union's most active official in Carbon County. Working closely with Radalj, Frank Bonacci was the union's most dedicated local advocate.

After the 1922 strike, Bonacci was blacklisted for his union activities. Leaving his family at home in Spring Glen, he traveled to Sublette, Wyoming, where he found work. Later, he returned to Utah and worked for the Utah Railway Company. By November 1923, Bonacci was back in the Storrs coal mines in Spring Canyon and was active as financial secretary for the UMWA local 1984 in Helper. One of his primary responsibilities as financial secretary was to contact those who had been members of the union and collect dues. The Globe Inspection Company labor spy in Castle Gate was among those contacted by Frank Bonacci and Tony Radalj. The visit by Bonacci and Radalj was described in the inspector's report dated February 18, 1925.

> These fellows told me the purpose of their call was to enlist my aid in this organization campaign and also urged that I pay up my back dues and get in good standing. I complied with the latter request, they placing me on the good standing list again for an initiation fee of $1.00, plus $2.50 for dues, a total of $3.50. I decided this was the best move to be made if I hoped to be able to keep in touch with the union end of it and in position to submit first hand information.
>
> I told these delegates I didn't consider it adviseable to take any active part in this camp, so far as organization work is concerned as it would simply mean the loss of my job and therefore can only speak for myself. During our talk they made the statement they have been making the rounds of the camps for sometime and have met with success in lining up new members, so that they are now quite well represented in all the camps. Said they hadn't experienced a great deal of difficulty in the Spring Canyon district, with the exception of one camp—Storrs, where they only have a few members. They made special mention of the camp of Rolapp, making the claim that the men there are lined up in good shape and is the strongest union camp in the state right now. Said they wanted to make a few calls before leaving Castle Gate and I know they called at three different places. From here they went to Rolapp and when leaving told me I could expect them back in Castle Gate within the next few days.[7]

The organizational work progressed through Bonacci's efforts. In his report on February 23, 1925, Inspector S.O. 2 noted Frank Bonacci had purchased a new car, which enabled him to visit most of the Carbon County camps under the cover of darkness and to sign up, within two weeks, nearly 200 new members.[8]

The dedicated work by Bonacci was enhanced by rumors of wage cuts; if the cuts were implemented, the Castle Gate labor spy warned, "some protest would undoubtedly develop."[9] The coal operators responded to Bonacci's union activities in the traditional manner. He was fired from his job at the Storrs mine. Bonacci informed District 22 officials of his circumstances, and told them that because of his union activities, he had been blacklisted from every mine in the area. In his correspondence, Bonacci requested a regular salary. The union officials, sufficiently impressed with his work and with the favorable prospects for organization in Utah, responded with an offer of a $100 monthly salary, a figure approximately half the amount he earned as a coal miner.[10]

The rumored pay reductions proved true when wages at the Kenilworth mine were cut 20 percent on April 15, 1925. A strike vote was immediately taken, resulting in a vote of 187 to 14 in favor of a strike. The coal company officials argued that a wage cut and a reduction in the labor force would allow the mine to be operated more than the previous one or two days a week, thereby giving the retained miners much higher earnings. In addition, the wage reductions would bring the Utah scale more closely in line to wage scales in other nonunion areas, which had recently suffered wage cuts. The wage cut was to be accompanied by a 20 percent reduction in the rent for company houses and 30 percent reduction in the amount charged at company boarding houses.[11] Samuel King described the strike's beginning in a letter to District 22 Secretary James Morgan.

> When the company got ready to notify the men of the reduction in wages, they sent the fire bosses through the mine, and without the men indicating what their intentions were, the fire bosses reported that more than 65 percent of the men would accept the reduction. . . . the notice was given on the 14th and to go into effect on the 16th of the month. One of their employees, and considered a trusted one, Mr. Ed Fox who has been with the company for about nine years, and without conferring with any of the other men, simply turned in his keys and stated he would not accept the reduction. Leaving the mine he went home and cleaned up, not anticipating the feeling of the others, and when he stepped out on to the street, found groups of men protesting against the matter. The result was a meeting was held and he took the leadership, and stated that he for one did not propose to accept the reduction, and the question was put as to how the other men felt, and every man voted not to accept the reduction. Then a vote was taken as to whether they would all stand firm against the reduction, and they all agreed to do so. The result was

> that on the day the reduction was to go into effect not a solitary man reported for duty.[12]

Nearly the entire work force of 300 to 400 men joined the strike, and the mine was shut down completely.[13] The coal company made no effort to bring in strikebreakers to operate the mine. Six weeks later, when market conditions improved sufficiently to justify reopening the mine, the old wage scale was restored and the men returned to work. Although rumors blossomed that other Carbon County mines would post reduced wage scales, they did not follow suit. This was, according to the coal miners, because of the determined stand of the Kenilworth miners.[14]

Because the wage cuts were confined to Kenilworth, a countywide strike, such as that in 1922, did not develop. It is doubtful whether the UMWA would have been any more successful than in 1922; however, the resistance at Kenilworth served as ample warning if wage reductions were made in other camps.

The apparent union strength did not persist, and nine months later Kenilworth officials laid off all the men who had participated in the strike. Union sympathizers comforted each other noting that "company officials don't know who they are hiring in their places and there have been almost as many union men hired as was laid off."[15]

Following the Kenilworth strike, Frank Bonacci continued to collect dues from the Utah miners. In January 1926, the Helper local union treasury showed a balance of nearly $600, and prospects were excellent for continued collections. According to the Globe Company inspector at Helper, Bonacci was optimistic about the future of the UMWA in Utah. In a recent district election, Tony Radalj had been reelected international board member from District 22, and Bonacci was sure this would mean continued help for Utah. In addition, Martin Cahill had been reelected District 22 president; and, in a letter thanking Bonacci for his support, Cahill stated that he could be depended upon to provide Utah with help whenever it was needed. The letter noted plans for a concerted organizing campaign to be initiated in Utah during the coming spring and summer, which would "put the Utah coal camps on the union map."[16]

The organizational effort began in March 1926 when Frank Bonacci arranged for the film *The Growth of American Labor* to be shown free of charge in Helper and Price. Although the film was shown under the auspices of the Organized Labor Bodies of Utah with no direct reference to UMWA sponsorship, it was clear that the aim of the film was to aid the organizational drive in Carbon County. On March 27, 1926, a Globe Company inspector assessed the role of the film and Bonacci's acumen in its use.

> Bonacci . . . has resorted to another scheme to stir up union sympathy among the coal miners. Realizing the difficulty in getting to each man individually for an opportunity to explain unionism, Bonacci decided to put on a moving picture at one of the show houses and which he believes will bring out the points more forcibly than he can do through talks. He first tried to get the Paramount house on Main street in Helper. The proprietor told him he was duty bound to show pictures that come in regularly every day but that he could put this picture on in addition, or as an added attraction. This was not satisfactory to Bonacci as he said he was afraid it would lengthen the program to such an extent that it would become tiresome and the audience not in a frame of mind to thoroughly appreciate the full meaning of the picture. Of course, there was to be no additional charge and the proprietor of the Paramount said that if he wished to have this picture put on as an added attraction it would be done free. This was not satisfactory to Bonacci and as the proprietor of the Paramount also has control of the Liberty Hall in Helper, Bonacci made arrangement with him for the use of that place at a charge of $15.00 and the picture is to be shown there on Tuesday evening, March 30th and the same picture shown in the City Hall at Price on Monday evening, March 29th. Both shows will be absolutely free; the union footing the bills. Hand bills have already been scattered around both towns advertising the shows.[17]

The use of film as a medium to promote unionization made a lasting impression on those who saw *The Growth of American Labor*. Previously, the union message had been whispered at clandestine meetings held in isolated locations by a few discontented men, who nervously looked over their shoulders in fear of company spotters and bulls. Now the union message blazed across the cinema screen with a new respectability, and perhaps faddishness, for organized labor. For those who were pro-union, or at least willing to consider the message of organized labor, the film gave unions a legitimacy that they were so often denied in the public press. The film helped Carbon County's coal workers understand more clearly their relationship with workers around the country and to realize that the union could be a reasonable and desirable American institution.

Bonacci kept in close contact with UMWA officials in Wyoming regarding the promised organizers. Reportedly, he expressed his inability to carry on the organzational work single-handedly and expressed the critical need for an American organizer, preferably someone like William Houston, who had played an important role in 1922.[18]

Instead of Houston, an organizer named Mooney was sent to Utah. He and Bonacci developed and implemented a plan whereby union members would be contacted in the various camps and would be asked to supply the names and addresses of men who could be considered good prospects for union membership. In an effort to sign them into the union, the organizers would approach these men in the evenings at their homes.[19] Martin Cahill, District 22 president, also visited the Utah camps in an effort to assess the situation, and the dutiful Globe Company inspectors reported the details of Cahill's visit to the coal operators.[20]

Union officials had picked an inopportune time for the organizational drive. Despite an apparent interest in the union by many Utah miners, the depressed state of affairs was a weak foundation on which to build a union organization. Mines were working only one or two days a week; and many miners, especially those with families, were leaving the camps in search of other employment. Another factor that hindered union development and solidarity was the antagonism between native-born and foreign-born miners. Illustrative of this problem was the lack of success by the Italian-born Bonacci with the American miners and a statement attributed to an Italian miner who declared, "The foreigners are not going to be the goats again and unless the Americans take the lead, there will be no strike."[21]

No labor difficulties occurred during 1926 with the exception of a minor incident at Rains where the miners went out on strike in March because they had not been paid for nearly three months.[22] There is no evidence that the strike at Rains was sponsored or supported by the UMWA, and apparently it lasted only a few days.

Throughout 1927, the union quietly continued its work in the Carbon coal fields under the direction of Frank Bonacci, who remained on the union payroll, and Tony Radalj, who made frequent visits to Utah from Wyoming. Local union officials were reported to be satisfied with the progress being made and confident that if the coal operators tried to reduce wages, 75 percent of the Utah miners would strike.[23] At the District 22 convention, held in Cheyenne in mid–March, delegates voted to aid in further unionizing efforts in Utah.[24]

Although not officially recognized by the Utah coal operators, the UMWA did prevent proposed wage cuts in 1922 and at Kenilworth in 1925. The union was able to do this largely because of the operators' fear of strikers. Nevertheless, the coal companies periodically tested the resistance of miners to wage reductions; in 1928, coal operators found that union strength and economic conditions had sufficiently deteriorated to allow a cut in wages at the Consumers mine in the Gordon Creek area. The new wage scale reduced day wages from $7.50 to $6.00, and coal mined on contract was reduced from seventy-nine to sixty-five cents a ton. Some fifty miners responded to the 20 percent pay cut by going out on strike.[25] However, the strike was short-lived and the new scale remained in effect.

Although the strike in Consumers was not called by the union, the lack of unity exemplified by it and the declining membership in Utah led District 22 officials to dismiss Frank Bonacci as field worker in Utah and to give full responsibility for the area to Tony Radalj.[26] In a letter to John L. Lewis, James Morgan requested that Tony Radalj be allowed to continue to work with the two Utah locals rather than abandon the UMWA effort in Utah.[27] Bonacci returned to the coal mines, working at the Maple Creek mine from 1929 to 1931 and then transferring to the Mutual mine. Early in 1933, he left Mutual for the Little Standard mine in Spring Canyon and worked there until he was rehired as a UMWA organizer in June 1933.[28]

John L. Lewis was not as optimistic regarding Radalj's chances for success in Utah. In a letter to Morgan, the national president made a viciously critical assessment of Radalj and countermanded his assignment.

> I had recently ordered Radalj to remain out of Utah, because he has been running down there at frequent intervals for several years, at very great expense to the International Union, and I was never able to discern that he had accomplished anything. His expense accounts for train fare and automobile hire, in connection with Utah, have, in fact, been enormous. As nearly as I can understand from questioning Radalj, he hires a motor car to haul him up to some camp at the head of a can[y]on, holds a whispered conversation in a secluded spot with two or three individuals, mounts his motor car and departs for another can[y]on. He asserts that he cannot, with safety, remain in any of the coal camps. Under such circumstances, I am extremely doubtful that he will ever increase the membership in Utah by a single individual.
>
> . . . You know, I am sure, that I want to aid and help the District . . . in every possible way and co-operate with them to the fullest extent in the Utah situation, just so long as there is any hope of getting reward for the money expended in that field. Radalj's activities in that territory for the last three and one-half years have depressed me very greatly.[29]

At Lewis's request, Martin Cahill, president of District 22, and George Young, district vice-president, visited the Utah coal camps in early

May 1929. They found the Utah miners extremely disappointed and dissastisfied with conditions. The Wyoming leaders found the hard feelings between American and foreign miners the greatest obstacle to establishing an effective union organization in Utah. In a letter to Lewis they reported:

> In our opinion, if the Native or American speaking Miner would take an active part or the lead, the Foreign speaking Miner would follow. There is quite a bitter feeling existing between the American and Foreign Miner, and on account of this feeling it is doubtful, in our opinion, if anything could be done along the lines of organization unless the American Miner would take the lead. In order to find out just what position the American Miner would take, a solicitation of these Miners should be made by men that could visit their homes, and find out just what stand they would take to get the State Unionized. We also believe that if the American Miner would take an active part, the state could be orginazed or partly orginazed without a strike. We believe that this information can be gotten at a possible expense of approximately $3,000.00 and the time required will be from sixty to eighty days. This is peculiar field and would require two men especially traimed along certain lines, in order to collect the information very essential before any more can be made to organization. We also advise that this investigation be made this summer.[30]

Lewis replied to Cahill that he was sorry to learn the bitter feelings and prejudices and acknowledged that it did create great difficulties in organizational work. In response to the request for $3,000 to survey American miners in Carbon County regarding their position towards the union, Lewis expressed his belief that the International Executive Board would not be willing to authorize such an expenditure.[31] At the meeting of the International Executive Board, held September 18, 1929, the board unanimously endorsed the reply made by Lewis.[32]

With union influence at its lowest point during the late 1920s and early 1930s, the coal miners turned to violence to protest long-felt injustices. When one of the Spring Canyon mines was worked on Labor Day in 1928, an unknown miner or group of miners sought to sabotage the operation by putting a rock in the switch for the tram and by removing the springs from the switch stand on the bridge at the tipple; either case would have caused a derailment. But before a trip of cars passed over the tram, company employees discovered the sabotage.[33]

Coal operators continued to seek wage reductions as the depression deepened in the early 1930s and, in one instance, miners at the Sweet mine struck against an 8 percent wage cut. On October 1, 1931, wages at the mine were again reduced from $6.00 to $5.50 a day, and fifty miners went out on strike in protest. The strikers, according to newspaper accounts, decorated their cars with placards protesting the company's actions and paraded the streets of Price.[34] They sought support from other miners by declaring if the Sweet mine officials were allowed to cut wages at the beginning of what most felt would be a hard winter in the depressed coal area, other coal companies would follow suit. The strike was short-lived, and the wage cuts remained in effect.

Throughout the 1920s, coal mining in the eastern Utah coal fields was in a depressed state, especially compared to the prosperity of the two earlier decades. In addition to the wage cuts that were imposed after 1927, the mines worked only one or two days a week. Many men left the coal fields, and those who remained competed with one another for jobs. The competition between American and foreign-born miners and the general spirit of intolerance were apparent in the investigation by the District 22 officials in 1929. Many American coal miners perhaps agreed with a letter written by Mrs. J. E. Tadlock of Murray to Governor George Dern regarding a situation at Bingham.

> We are a family that have never asked any help in the past from anyone. But when there is no work for men there has got to be something done.
>
> I am calling on you for help. We have to have a chance to earn what we eat and wear, and fuel to keep us warm.
>
> We have at home two able bodied men willing to work but can't get it. They are good miners. In Bingham we have the mines filled with Bohunks, Austrains and Greeks, all taking the white mens jobs.
>
> Now something could be done in Regards to these foreighners.
>
> Even if you did not send them back to their own country, they could be layed off and put the white men to work.
>
> If they would do that the white man could survive.[35]

Both foreign-born and American miners found little sympathy among coal company officials for their plight. In a letter to Governor George H. Dern in 1930, an Austrian-born miner, Steve Zamof, explained his adverse situation and the lack of concern by an official at the Columbia mine.

> I was previously employed at the Columbia Mine and was layed off work June 5th account of reducing the force and have been there several times, trying to seek employment. On my last visit at Columbia explained to Mr. Harvey, the condition of my family and pleaded with

> him to furnish me employment as I had four children whom were hungry and he refused to give me work and said he did not care if I had ten children.[36]

Three years later, the situation was much the same; an American-born miner, Hugh McMillian, described to Governor Henry H. Blood the inflexibility of the United States Fuel Company regarding housing in the town of Heiner.

> I am writing you in regards to the house proposition in this town. There is about 15 empty houses here that are just being torn to pieces and I have been renting here for the past year and being unable to get work I have not been able to pay the rent so they sent me a letter which I am mailing to you so you can see for yourself what they are doing. I have 9 children and my baby is sick and they said if I wasn't out in 3 days they would have to come and throw us out in the street. I have taken good care of the house ever since I moved into it but they said they would rather have them setting empty than to have people living in them who could not pay their rent. There are four or five families here who received the same kind of letters yesterday. I have worked in the coal mines here for the past 25 years and it doesn't seem fair to me. . . . I don't want anything for nothing and am perfectly willing to work to pay my rent and support my family but it is impossible to get work of any kind around here.[37]

For other Utah coal miners, whose situations were similar to those of Steve Zamof and Hugh McMillian, the election of Franklin D. Roosevelt in 1932 provided a hope. They were not disappointed. The provisions of Section 7a of the National Industrial Recovery Act (NIRA) ended the years of conflict over union recognition. By May the question for most coal miners was no longer: would the coal companies agree to recognize their union, but rather, which of two contenders—the National Miners Union or the United Mine Workers of America—would be selected to champion the coal miners' New Deal?

Chapter X

The Battle for Control, 1933

John L. Lewis is credited with immediately realizing labor's potential under Section 7a of the National Industrial Recovery Act and with carrying out the organizing campaign that dramatically reversed the downhill course American labor had followed since 1919.[1]

In assessing the significance of the National Industrial Recovery Act, Lewis declared, "From the standpoint of human welfare and economic freedom, we are convinced that there has been no legal instrument comparable with it since President Lincoln's Emancipation Proclamation."[2] At a later point in his address to the delegates of the 1934 UMWA International Convention, Lewis outlined the vigorous nationwide organizing campaign that had been intiated under his direction.

> Coincident with the signing of the act by the President, the United Mine Workers of America conducted a vigorous organizing campaign in all the mining Districts of the United States. Systematic plans were laid out, meetings were addressed by able Field Representatives, and the enrollment of new members took place upon an unprecedented scale. It was easily demonstrated that the mine workers employed in the non-Union areas of the mining industry would enthusiastically join the UMWA if they were privileged to do so. Local Unions were established, local Officers selected and installed, supplies furnished the Local Unions, and in less than 30 days from the signing of the act by the President, the complete organization of the bituminous industry was effectuated.[3]

Although Lewis deserves recognition for his leadership of the nationwide organization campaign, it is apparent that the UMWA organizational drive in Utah was as much in response to local opportunities and threats as it was part of any systematic nationwide plan. The opportunity for the UMWA came in May 1933 when miners at Standardville struck over a reduction in wages; the threat to the UMWA came from the rival National Miners Union (NMU), which entered the Utah coal fields several weeks before the proposed NIRA was sent to Congress.[4]

In early May, the Standard Coal Company cut wages of men working in its Standardville mine in Spring Canyon. The miners maintained that the reductions violated company promises to the Roosevelt administration and went out on strike.[5] John L. Lewis was immediately apprised of the Standardville situation by a telegram from George Cole, president of District 22.[6] Cole saw the strike as an excellent opportunity to organize the Utah coal fields and sought support from Lewis to send a sufficient number of organizers to accomplish the task. The reply from Lewis was noncommittal. "It is practicable to take advantage of any situation in Utah which may lead to organization but suggest it is unwise to assume any financial liability under present circumstances."[7]

In his next correspondence to President Lewis, George Cole, reflecting the influence of the lukewarm response, explained that his intention was to send Tony Radalj to Utah to determine the cause of the strike and the feelings of the Utah miners about unionization. Cole concluded: "If the report is favorable then we will go down there and do the best we can; I realize that we will have to be careful down there because we don't have any money to get us out of trouble if we should get into any. It's hell to be poor John."[8]

Lewis concurred with the plan to send Radalj to Utah and again cautioned against incurring the large financial obligations of "strike benefits, legal expenses and other things which usually come from a strike in a non-union section." Looking beyond the immediate situation at Standardville, Lewis saw the organization of Utah more likely after the passage of national labor legislation that would give "to labor the right to organize and to collectively bargain through representatives of their own choosing."[9]

In less than two weeks, John L. Lewis changed his mind about an organizational campaign in Utah. On May 27, 1933, three weeks before the NIRA was signed by President Franklin Roosevelt, International Representative Nicholas Fontecchio was charged with organizing the Utah coal miners. Fontecchio was supposed to cooperate with the District 22 officers, but Lewis made it clear that Tony Radalj, also an international board member, who had been responsible for Utah during much of the 1920s, was to work under the direction of Fontecchio.[10] Frank Bonacci was rehired as an organizer, and Utah quickly became a major objective in the UMWA campaign to complete the organization of the nation's coal mines. Throughout the summer and fall of 1933, John L. Lewis was informed almost daily of the progress in Utah.

In the meantime, Tony Radalj had visited Utah as requested by District 22 President George Cole. His report to John L. Lewis made no mention of a pronounced sentiment for UMWA organization. He "found conditions in the Utah fields very quiet, mines operating very slowly and with materially reduced and skeleton forces." Continuing his report, Radalj noted the presence of the National Miners Union and mentioned that it had been active "in the different Utah camps throughout the past winter."[11]

The reason for John L. Lewis's decision to begin the Utah organization campaign before the passage of the NIRA is difficult to identify. Perhaps by late May Lewis was able to gauge what the congressional and presidential responses to the provisions in Section 7a would be. Perhaps on this basis, he set the organizational machinery in motion in anticipation of the bill's passage. A second possibility is that Lewis, though concerned that Congress and the president give labor its "Emancipation Proclamation," came to realize that the sentiments of the country's nonunion miners strongly favored unionization. He, therefore, undertook the organization of the nonunion areas with the conviction that he would succeed even without the provisions of Section 7a. Although the Radalj report on the National Miners Union was submitted after the decision by Lewis to send Fontecchio to organize Utah, a third explanation is that Lewis was aware of NMU activities and concluded that only a quick and intensive campaign would crush the NMU before it organized the Utah coal fields.[12]

The NMU was formed in 1928 by disenchanted UMWA members who believed they had been sold out by John L. Lewis and the UMWA during the 1927–28 strike in Pennsylvania and Ohio. In September 1928, the first convention of the NMU, held at Pittsburgh, Pennsylvania, was broken up by UMWA sympathizers. Yet discontent with the parent union flourished. The NMU proved much more militant than the UMWA and, despite a shortage of funds, initiated strikes in Illinois in 1929; in Ohio, Pennsylvania, and West Virginia in 1931 and again in 1932.[13]

The five fundamental principles of the NMU indicate the radical nature of the organization. The union existed to organize and lead all miners irrespective of race, color, nationality, age, sex, or employment status in their struggle for better working and living conditions against capitalist mine owners. The union's ultimate aim was to overthrow capitalism and replace it with socialism. Since the government was the tool of capitalists, it was essential for the union to fight for the workers' basic political rights and to support workers against their exploiters through politics. The NMU supported all workers, not just miners. The union sought to create a united front of all miners and of the rank and file in the reformist trade unions for "common action . . . against the common enemy, the coal operators, the labor bureaucrats and their government."[14]

Despite Tony Radalj's report that the NMU had been active in the Carbon County area during the past winter, the union's presence was not formally announced until May 28, 1933, at an organizational meeting in Helper. Charles Weatherbee presided over a meeting of 500 people, most of whom were foreign-born.[15] Several speakers, including Paul Crouch, addressed the assembly. They rehearsed the typical miners' grievances: low wages, cheating at the weighing scales, abuses of the company town system, and a lack of concern for safety in the mines. The NMU speakers also attacked the Hoover and Roosevelt administrations and the UMWA. At the conclusion of the meeting, Weatherbee announced that five local miners had agreed to work with the NMU organizers in their respective areas. Among those identified were Jack Hall of Peerless and Tony Bonacci of Helper (no relation to Frank Bonacci of the UMWA).[16]

Weatherbee also outlined the demands that the NMU would seek: recognition of elected grievance committees and the right to organize; checkweighmen who would be elected by the miners; a 30 percent reduction in rent and board charged for company housing; wage increases commensurate with the increased cost of living; a regular payday twice a month and immediate payment of all back wages; pay for all "dead work," which included timbering, laying track, and other work for which miners working under contract had not previously been paid; and the right to live anywhere the miner chose.[17]

The NMU organizational drive appeared to have been carefully conceived. Four days after the May 28 organizational meeting, the union began publishing the *Carbon County Miner*.[18] The weekly newspaper was published at the *Helper Journal* office and included advertisements from many local business establishments. The paper served as an effective means to inform NMU members of activities, meetings, and new developments.

Headquarters for the NMU in Carbon County was established at Millerich Hall in Spring Glen, a small community just south of Helper.

The building was a rectangular, one-story structure constructed by striking coal miners in 1922 and operated by Martin Millerich, a South Slav and ex-coal miner. The hall was opened daily for those interested in joining the union or for meeting with officials and other members. Millerich Hall was also used for some of the Sunday meetings and Saturday night dances sponsored by the NMU. In addition to Millerich Hall, Sunday meetings were also held in the Helper Park after baseball games (played by teams of the semi-professional Utah Industrial League) were over.[19] At times, attendance at these meetings ranged between 500 and 1,000.

According to the *Carbon County Miner*, Utah miners flocked to join the NMU. On June 14 the paper reported: "Hundreds of miners, from every mine in Carbon County have joined . . . during the past three weeks. Applications are coming in at the rate of fifty per day."[20] By June 28, a month after the organizational campaign began, the NMU membership was reported at 1,000.

The NMU also organized a Women's Auxiliary at Millerich Hall on June 7. Sylvia Crouch, wife of organizer Paul Crouch, explained the importance of organizing the women as well as the men.

> Wives and sisters of the miners of Carbon County are learning that they must fight side by side with the miners. If the women are not organized and do not understand the struggles of the men they will find themselves the victims of all kinds of gossip and lies circulated by the companies and their agents. They try to fool the women and attempt to cause them to "nag" at their husbands during strikes, etc. The companies and their stool pigeons, however, will be doomed to disappointment because the Women's Auxiliary is on the job. And when a strike begins the women will be found on the picket lines to prevent any scab from taking the places of their husbands and brothers.[21]

Members of the Women's Auxiliary also arranged for the Sunday picnics that brought many new members to the NMU. A Youth Section was organized on June 22 to educate both the young men and girls of the coal camps and to provide social and sports activities. The Youth Section met regularly at the local roller-skating rink.

With the Thursday publication of the *Carbon County Miner*, the Thursday evening meetings of the Women's Auxiliary, the Friday evening meetings of the Youth Section, the Saturday night dances at Millerich Hall, and the Sunday picnics and meetings, NMU activities dominated the lives of its members. This, in turn, led to a great solidarity within the ranks of the Carbon County members of the NMU.

Despite assertions that locals of the NMU were being established in all Carbon County mining camps, the organizing activities of the NMU seemed confined to those camps whose social and economic hub was the city of Helper. Accordingly, the NMU was most effective in the Gordon Creek, Spring Canyon, Kenilworth, and Castle Gate areas and saw little or no success in the towns of Scofield, Clear Creek, Hiawatha, Wattis, Sunnyside, and Columbia. In these communities, the UMWA was able to organize with little interference from the NMU when the national organizers arrived in mid–June to begin the UMWA campaign.

In a letter to International Vice President Philip Murray, Nicholas Fontecchio reviewed his warm welcome and immediate success in Carbon County, noting he had met with the miners, coal operators, and leading citizens and was well along in his mission to establish local unions throughout the area.[22]

Fontecchio established a two-room office in the Silvagni building in Price and, with his corps of District 22 officials and local workers, organized seventeen local unions within three months: Local No. 5807, Wattis; No. 5806, Standardville; No. 5916, Castle Gate; No. 5861, Kenilworth; No. 6089, Columbia; No. 6022, Peerless; No. 6204, Scofield; No. 6210, Spring Canyon; No. 6244, Sunnyside; No. 6072, Consumers; No. 6363, Hiawatha; No. 6412, Rolapp; No. 6508, Latuda; No. 6510, Mohrland; No. 6509, Salina; No. 6511, Sweet; and No. 6567, National.[23]

With two unions in the Utah coal fields competing for members, they soon tried to discredit each other in the eyes of the miners. The NMU branded the UMWA as strikebreakers, as agents of the coal operators, and as a company union and argued that Utah miners had been sold out by the UMWA during the 1922 strike. The *Carbon County Miner* explained how miners at Peerless and Standardville were forced to join UMWA or face losing their jobs. At Kenilworth, the superintendent reportedly provided transportation to a UMWA meeting in Price and promised ice cream to the children of those miners who attended.[24]

Members of the UMWA declared that they were constantly subject to threats, intimidations, and insults by NMU organizers and members.[25] The UMWA officials countered that the NMU was a communist organization. In a pamphlet entitled "The New Deal: For the Coal Mining Industry," Nicholas Fontecchio wrote:

> Please do not mistake the United Mine Workers of America for the National Miners' Union which is now in your midst blowing their trumphets of discord only to satisfy an irresponsible and disgruntled

element. There is a great deal of difference between the two: The United Mine Workers of America is purely an American institution, working under American ideas and ideals and believing in solving its problems in line with the American laws and policies. The National Miners' Union, on the contrary, is not an American labor organization, nor is it affiliated with the American Federation of Labor. Their policies are dictated from Moscow, Soviet Russia. This so-called miners' union is a menace to the coal miners, to the coal industry, and to our Nation. Their conduct for a few years past should be intelligently analyzed. The United Mine Workers of America knows only too well they have been responsible for much trouble in the coal industry. We are not ready to subscribe to the philosophies of Russia. The United Mine Workers of America believes that there is room for only one flag in the United States and that is the stars and stripes, and by the same token believes that there is room for only one miners' union and that is the United Mine Workers of America.[26]

Resolutions denouncing the NMU officials as agents for the Communist party were passed at local meetings of the UMWA and were sent to Governor Henry Blood.[27]

Representatives of the NMU did not deny that there might be communist influence within the union. The philosophy of the union, as outlined earlier, was reminiscent of other socialist, if not pro-communist, groups. The *Carbon County Miner* responded to charges of communist influence by stating: "Yes, we HAVE members of the Communist party. We ALSO have members of the Democratic, Socialist and Republican parties, as well as a far larger number of miners not connected with ANY party."[28] Continuing the same line a month later, the *Carbon County Miner* acknowledged that Communist party members were elected to union positions by their fellow workers "because they have shown themselves militant fighters for better wages and conditions." The article denied that orders for the NMU came from Moscow. "How many sane people," the article asked, "seriously believe that 'Moscow' decided that the Carbon County miners should demand a checkweighman, regular pay days, the right to trade anywhere they wish, 55 cents per ton of coal, etc."[29] In a meeting held September 4, Charles Weatherbee declared that the NMU was not a communist organization but suggested that "perhaps if it was it would be a still better organization."[30]

While NMU leaders were careful to emphasize the immediate concerns of the Carbon County coal miners and to downplay the far-reaching objectives of the Communist party, further evidence suggests that the

NMU organizers were communists and that the Carbon County National Miners Union was linked to a nationwide network of communist-directed organizations that had strong grassroots support from those devastated by the depression. At the first meeting of the NMU in Millerich Hall, a red banner with hammer, sickle, and sheaf of wheat decorated the room; and in later years, Paul Crouch would openly declare his membership in the Communist party.[31]

The Carbon County Unit (Section 3, District 19) of the Communist party was organized shortly after the NMU organizers entered the county. Local communists identified themselves in a handbill entitled "Who are the Communists?"

> The Communists are ALWAYS on the side of the workers in every struggle for better conditions. The Communists throw their full support to every union—to every organization—*which fights for better conditions of the working class.* The Communists ARE workers who know that the working class can gain nothing without struggle. But we must do MORE than fight for immediate demands. We must understand that the government is controlled by the mine operators and others of the boss class, and is used by them against the workers. . . . The Communists believe that the means of production should belong to those who use them, who produce ALL wealth—the working class.[32]

Though leaders generally tried to disassociate the NMU or local Communist party from the Soviet Union, in one article published in the *Carbon County Miner* a glowing comparison was made between coal miners in Russia and those in the United States. In the Soviet Union, the number of coal miners had doubled in the past four years and wages increased 210 percent. Soviet miners reportedly enjoyed many benefits their American counterparts did not: paid vacations, free medical treatment, dining rooms near the mines open day and night so the miners could get hot meals before or after their shifts; and a workday of six hours.[33]

The NMU was branded as communistic also because of its vocal criticism of the policies and personnel of the Roosevelt administration. In the organizational meeting of May 28, Paul Crouch reportedly declared that "the R.F.C. [Reconstruction Finance Corporation] was giving people help only in such a way as to keep them partially satisfied and to divert the attention from things that were important."[34] Charles Weatherbee, skeptical that Franklin D. Roosevelt and Frances Perkins, secretary of labor, were really concerned about the poor people, claimed that when millionaires were appointed "to do things for the poor people it doesn't look sincere."[35] The Civil Conservation Corps was seen as a military endeavor "to get the best young men of the country . . . who could pass the physical examination, and harden them up, get them accustomed to military discipline so that they could put guns in their hands at any time and put them to fighting any place they wanted and at any time."[36]

Overreacting to the NMU criticism of Roosevelt and the New Deal programs, State Director of Public Welfare Robert H. Hinckley threatened that "unless the vicious attacks made against the President and National Recovery Act in Carbon County were discontinued, government action would be taken."[37] The implication was clear—unless the NMU criticisms stopped, federal relief funds to the entire county might be cut off. The NMU had a field day in responding to Hinckley's declaration. The *Carbon County Miner* righteously proclaimed:

> This is probably the most open move in the direction of a Fascist dictatorship ever made by a public official in this country. It would completely destroy all freedom of speech and press, and prevent the workers from fighting any starvation code which had the O.K. of the NRA administrator. Exercise of the "constitutional right" of criticism by individuals of actions of the president would be punished by starvation of the entire families of unemployed throughout the county.[38]

Because of the reported communist influence and the attacks against President Roosevelt and his programs, the NMU was denied permission to participate in the Fourth-of-July parade in Price. Traditionally, the Fourth-of-July celebration alternated between Helper and Price. In 1933, it was to be held in Price, and Mayor Rolla West asked the Price post of the American Legion to sponsor the celebration. When NMU officials applied to the American Legion Post for permission to join the parade, they were refused, despite an uncontested announcement by UMWA officials that delegations from their locals in Carbon County would participate in the parade and celebration. Later, because of the unfairness in allowing one union to parade while denying the right to another, permission for the UMWA to march in the parade was rescinded.[39]

The issue left bitter feelings among county residents, especially when parents of members of the Carbon High School band declared, "If we are bolsheviks, so are our children and we will not allow them to parade and play in the Price [Carbon County High School] band on the Fourth."[40] The band had recently returned from the Chicago World's Fair with first place honors in the band competition.

A separate Fourth-of-July celebration was sponsored in Helper by the NMU with Charles Weatherbee as parade marshal and Charles Guynn the main speaker. More than 4,000 people participated in the Helper celebration.[41]

While Guynn and other speakers in Helper compared the Revolution of 1776 with the present need for miners and other members of the working class to carry out their own revolution for economic independence, in Price Mayor Rolla West declared that it was time for the American people to rededicate themselves to the principles set forth in the Declaration of Independence and in the Constitution of the United States rather than turn to "communism, bolshevism, or any other policy that poisons the human mind."[42]

Helper Mayor Frank R. Porter approached the problem of the NMU-UMWA conflict much differently than did Mayor Rolla West. In an article written by Porter and published in the *Salt Lake Tribune* a few days after the Fourth-of-July controversy, the Helper mayor declared himself neutral regarding the two labor organizations. He noted that it would be up to the coal miners and mine operators to decide which union would be in control. In contrast to the policies of Rolla West, Porter explained that he would not "use [his] official capacity to intimidate or deny certain rights and privileges allowed . . . by the constitution . . . so long as there is no trouble or rioting."[43]

The Independence Day episode did prove advantageous to the NMU in the north Carbon area. The local animosities between Helper and Price were magnified because Helper residents saw the handling of the NMU by West and the Price American Legion as an insult to their own community, and, unintentionally, the NMU became their champion against the dictates of Price. The hard feelings between the two communities were expressed in several ways: through the Carbon High School band issue, criticism of Rolla West by the *Helper Journal*, and through social ostracism (a truck that appeared in Price to advertise a dance in Helper at the Rainbow Gardens was ordered "to get out of town and stay out").[44]

The NMU continued to hold frequent meetings, but its policy in Carbon County was not one of rhetoric alone. Following the same course as the national union throughout its stormy five-year history, the local NMU organization was determined to strike at the first opportunity. At a July 9 meeting of the NMU held in the Helper City Park, the 900 people in attendance were informed of plans for a general strike to begin on Labor Day.[45] Members of the NMU working at Mutual were not content to wait for the general strike, and on August 2 they walked out in an attempt to gain recognition of the NMU. The strike was successful, and, in addition to recognition of the NMU, the miners won the right for a checkweighman to be elected by the miners, drivers to be paid by the coal company instead of by miners, regular paydays, the right for the miners to select their own doctor, and a promise that the discharged miners would be reinstated and that preference would be given to hiring unemployed miners living in Mutual rather than to recruiting miners, especially anti-union men, from outside the camp.

The agreement was not permanent, and shortly after the men returned to work, Superintendent Oliver Sutch was charged with violating the terms of the settlement when he reportedly brought in nonunion men from a farming area and gave them employment instead of hiring miners living in Mutual in company houses. NMU leaders declared that it was "an effort to replace the NMU members so former conditions could be restored," and struck the Mutual mine for a second time.[46]

Superintendent Sutch saw the second strike in a different light, and he reported that the NMU was demanding that UMWA members and unaffiliated miners be discharged. The NMU requested that Sutch be removed from his position as superintendent but be allowed to name his successor. Sutch stated that the Mutual mine would be closed until it could be operated "under a fair and equitable code."[47] Despite his firm declaration, Sutch was overruled by his superiors. After a week-long strike, marked by mass picketing by the strikers, the NMU Women's Auxiliary, and the NMU Youth Section, the NMU demands, according to the *Carbon County Miner*, were met, and Superintendent Sutch left for a "long vacation."[48]

The success of the NMU at Mutual prompted miners at Spring Canyon to walk out on strike the evening of August 16. The strike quickly spread to the Gordon Creek area where miners at National voted to strike on Sunday, August 19, and those at Consumers mine and Sweet mine went out the next day.[49]

The demands of the Spring Canyon miners were more temperate than those of the Mutual miners and unexplainably did not refer to NMU recognition. The demands did include recognition of a representative mine committee elected by the miners, a checkweighmen elected and paid by the miners, regularly semimonthly paydays, time-and-a-half for work beyond eight hours a day, a 25 percent reduction in house rent for company houses, and fifteen hours notice of when the mines would be worked.[50]

According to reports by Spring Canyon Superintendent David Brown and Sheriff S. M. Bliss, NMU membership in the strike areas was less than that of the UMWA.[51]

	NMU	UMWA	Unaffiliated
Spring Canyon	110	110	(not given)
Consumers	27	70	15
Sweet	23	43	20
National	30	2	8
Total	190	225	43

The strong pickets maintained by the NMU soon led to a confrontation between the two unions. UMWA members, armed with pick handles, rifles, and shotguns, tried to break the picket lines and, according to statements by members of the NMU, attacked women pickets with fists and pick handles.[52]

The strike and threat of violence led local businessmen and coal mine owners to join ranks with the UMWA leaders in a unique alliance that requested Governor Henry Blood to send National Guard troops to Carbon County. In a telegram to the governor, Nicholas Fontecchio described the situation from the UMWA perspective. "Threats intimidations and insults prevents our members from working . . . our men are reluctant to work side by side with communists . . . unless communists are eliminated nothing but chaos will follow."[53]

Residents of Spring Canyon reported that the highway into town was continuously blocked by a truck and by strikers who jumped on automobile running boards and who stopped motorists from crossing picket lines. Men carried clubs and revolvers while women loaded their purses with red pepper to throw in the eyes of their antagonists. The pickets were maintained throughout the night, and complaints were registered by those who found sleep impossible because of the "noise and shouting."[54]

Governor Henry Blood responded to the many requests for the National Guard by calling a meeting of coal mine owners and managers, businessmen, state and local officials, and other interested individuals to discuss the issue. At the meeting, held in the governor's boardroom in the State Capitol Building, the forty to fifty men present were given the opportunity to express their feelings on sending the National Guard to the strike area. Most of those present wanted the National Guard called out, but Sheriff Marion Bliss, County Commissioner Bill Reed, and Price Mayor Rolla West felt the local situation did not warrant it.[55]

W. M. Knerr, head of the State Industrial Commission and advisor to Governor Charles Mabey during the 1922 strike, advised Governor Blood, "There is absolutely no reason at this time to even consider any proposal to send the state militia to this county."[56] Although a proclamation and executive order calling out the National Guard and declaring martial law in Carbon County was drafted, it was not signed.

Law enforcement was left to local authorities; and a few days following the Salt Lake meeting, Rolla West allegedly was asked by Governor Blood to organize and direct a force of local citizens in an effort to resolve the lawless situations created by the NMU strike. The thirty-eight-year-old West had lived in Price since 1910; he had served with the American forces in France during World War I. A builder and carpenter by trade, West was a longtime member of the United Brotherhood of Carpenters and Joiners. As a contractor, West hired only union members, paid union scale, and observed all trade rules. West enthusiastically supported the UMWA; he appeared at and participated in its public programs and Labor Day celebrations. As mayor of Price, he was aggressive in the establishment of a worker-relief program in the depression-torn county.[57] But the NMU found West to be one of its most vocal and active critics. Though himself a loyal union man, West could not tolerate the radicalism, anti-Americanism, lawlessness, and communism he found in the NMU.

As chief deputy sheriff in Carbon County, West was to be responsible only to the governor. Although this arrangement apparently left Sheriff Marion Bliss in a subordinate role, the two men worked together to recruit a force to carry out campaigns against the NMU strikers in Gordon Creek and Spring Canyon.[58]

With apparently little fear of the NMU strikers, West began his recruiting campaign in the heart of NMU territory with a meeting at the Standardville Amusement Hall. Nicholas Fontecchio and Frank Bonacci of the UMWA reportedly offered their services in recruiting deputies, and within twenty-four hours Rolla West boasted more than 100 volunteers.[59]

The force, made up almost exclusively of UMWA members, was unique in Utah's history. It marked the first time that members of one union were used as deputies in an effort to break a strike by another union. In addition to the deputies, who were armed with weapons loaned by the National Guard, two detachments of National Guard troops were sent to assist the local authorities. The first detachment, under the leadership of Lt. Col. Albert E. Wilfong, arrived the evening of Thursday, August 24, and the second detachment, a riot squad from the 145th Field Artillery and led by Lt. Harry A. Randall, arrived the next morning.[60]

The Gordon Creek camp at Consumers was regarded by local authorities as the most urgent problem. Here the strikers, most of whom were South Slavs, had blockaded the public road with railroad ties, mattresses, and a truck. The superintendent at Consumers, David Parmley, who was also chairman of the Carbon County commissioners, was especially despised by the Slavic miners. The deputies were instructed to report to the Carbon County Courthouse at 4:00 a.m. Saturday, August 26. Armed with the National Guard rifles and carrying the lunches they were instructed to bring, the deputies were loaded into state road trucks ("very slow, old white GMC World War I surplus vehicles") and transported to a section house four miles from Consumers.

Fearful of an ambush, scouts were sent up the ridges on both sides of the canyon while the main force waited. At noon a runner returned to report that men armed with rifles were guarding the road to Consumers. The scouts were unable to determine whether the men were NMU strikers, or highway patrolmen who had been sent to the area earlier to keep the road open, or a company of Consumers deputies under the command of Joe Bono. West concluded that they were probably NMU strikers and chose to wait for further word before moving his men closer to the camp. The men were already suffering because of the hot, shadeless terrain and because they had failed to bring water with them, mistakenly anticipating that it would be available from Gordon Creek.

Later in the afternoon, David Parmley came down to the camp. Although Parmley had eluded the armed strikers, he was followed down the road by "6 big Austrian women" who, as reported by Parmley, "threw me down, took my revolver, threw my shells away and held me down and peed on me."[61] A short time later, a dozen young NMU sympathizers followed the same route and were effectively handled by Castle Gate's Mormon bishop, Bill Staply.

> Bill went up to the biggest fellow who seemed to be taking the head and said, "What are you doing here?" The young man started to say something and Bill said "You fellows have been trying to tell us what to do all summer" and pow he knocked him down. Before this bunch could take to their heels Bill had knocked down two more of them and his partners didn't get in a lick. They called Bill the fighting Bishop after that.[62]

At sundown the deputies were finally loaded into trucks and transported to the entrance of Consumers. There they were organized into four companies and marched through the camp toward the mine entrance where 200 NMU members congregated. As the deputies came closer to the mine, the strikers quickly dispersed, some going inside the mine and others making their way north up the ravines, apparently headed for Spring Canyon.

With the camp secure, Joe Bono and his Consumers detachment were left to maintain order while the remainder of the deputies were loaded into the trucks for the return trip to Price. The first campaign was successful. Consumers was liberated and the only casualties were the humiliated David Parmley and the "3 young no-name-ovichs with fat lips awarded them by a Mormon bishop."[63]

The next day, Sunday, August 27, West and his men turned their attention to Spring Canyon and made plans for an expedition the next morning. This time the men were equipped with water-filled whiskey flasks, which fit comfortably in their hip pockets and prevented the men from "spinning cotton," as they had in Gordon Creek.[64] A three-pronged attack was devised with deputies marching down the canyon from Latuda and Standardville with companies from Kenilworth and Castle Gate going up Hard Scrabble Canyon, just north of Spring Canyon, and then crossing the ridge into that canyon above the town; and with deputies from Columbia and Hiawatha meeting deputies from Spring Canyon and old Peerless in the canyon below the town.[65]

The plan proved effective. With the use of tear-gas bombs, the NMU pickets—including men, women, and children—were surrounded by the deputies with little difficulty and were marched a half-mile to the Peerless mine where 210 men were loaded into trucks that transported them to Price. In Price, they were incarcerated in a bullpen at the ball park and in an empty Price business building.[66]

Throughout the day, virtual martial law was in effect. A carload of Italians on their way to a lodge meeting in Helper was detained by West when it was learned that one of the men was an Italian baker whose Spring Canyon bakery was a known meeting place for the NMU. Later, the baker's wife appeared to protest the detention of her husband and struck West in the face with her satchel, giving him a bloody nose, and then kicked him in the shins. Arrested and taken to a nearby automobile, the woman had the car door slammed on her legs by an overly zealous coal company clerk.[67] An Italian boy suffered a broken arm when one of the deputies reportedly tried to take the boy's rifle.[68] Harold Huff, president of the NMU local at Mutual, was arrested, handcuffed, placed in a boxcar, and then given the chance to leave the county and not come back.[69]

When NMU members in Helper learned of the action against their Spring Canyon comrades, they quickly organized a delegation and began to march up the canyon. The group was led by thirty to forty women, with two "very pretty young ladies" carrying American flags. They were met by deputies who arrested the women after they threw pepper in the deputies' eyes. Reportedly, the flag-carrying women were both slapped in the face and had the American flags torn from their hands. The women were transported to Price in state road trucks and then released that evening by officials who were unconcerned about how the women would return to their homes several miles away.[70] The *Helper Journal* concluded its report on the Spring Canyon raid by describing the events as tantamount to civil war.

> Miners have borne arms against fellow workers; school boys have hurled tear gas bombs into the picket ranks of their former chums, and a general disorder has resulted from throwing armed deputies into the picket lines which has resulted in much bitter feeling, ill will and enmity which will hold for many years to come.[71]

Two days after the Spring Canyon raid, Tuesday, August 29, six or seven truckloads of deputies led by Rolla West invaded Helper. After surrounding the Carbon Hotel, the deputies broke into a room occupied by Charles Guynn and Charles Weatherbee and placed them under arrest. Expecting much worse than arrest, the men were described as "very surprised and very scared. Their faces were an ashen yellowish-green."[72] It was not until the arrest warrants were read that the two NMU leaders realized that their destination was jail rather than a hanging tree. Their concern was real. After spending a short time in jail, along with Paul Crouch, who had been arrested while in Salt Lake City and transported back to Carbon County, they were released on $5,000 bail. A group of deputies who had participated in the arrests at Helper met in the Savoy Hotel in Price to formulate plans for recapturing the NMU leaders and for permanently disposing of them. Mayor Rolla West was invited to the meeting; and just as the men were ready to leave the room, he was able to persuade them that they would regret such violence.

> Just maybe there is one little thing that we haven't considered. . . . I think we are going to get away with a clean deal tonight, but what

> about tomorrow and the tomorrow after that. All of you fellows have church affiliations, you all know I don't. I can truthfully say that I could do away with these S.O.B.'s and never give the matter another thought. But perhaps every man should ask himself now, before we start out, "How is this deal going to effect me, my relation with my family and with my religion after I come back." Every man was quiet. No word was spoken.[73]

The strikers arrested during the Spring Canyon raid and held in the Price bullpen were almost forgotten. For the first twenty-four hours, they were not given food, water, or blankets. Finally, arrangements were made with a local restaurant to supply the men with food and coffee. This was supplemented with home-canned fruits, vegetables, and cured meats brought by wives and friends who visited the prisoners daily.[74]

A few of the prisoners were able to raise the $1,000 bail and were released from the bullpen after promising to return to work and "maintain the peace." However, approximately 200 remained in the bullpen. One hundred thirty of the prisoners were aliens, and deportation proceedings against them were initiated, though not carried out.[75]

The raids in Gordon Creek and Spring Canyon and the resulting arrests did not deter the NMU from continuing plans for a general strike to begin September 4. The strike call was directed to all miners—NMU members, UMWA members, and unorganized miners. The objectives of the general strike were outlined in the *Carbon County Miner* and on printed broadsides as follows: a six-hour day and five-day week; a minimum wage of one dollar per hour; fifty-five cents per ton for machine coal and ninety-five cents per ton for pillar coal; pay for all dead work such as laying track and timbering; checkweighmen to be elected and controlled by the miners; the right of miners to join any union of their choosing; the immediate reinstatement of all miners who had been fired because of the strike activity; and a conference between elected representatives of the miners and representatives of the companies to work out a wage scale for the Utah coal mines.[76]

The strike call coincided with the opening of school and NMU officials sought to include a school strike by the members of its Youth Section as part of the general strike. The purpose of the school strike was to demand the removal of teachers who had served as deputies during the late August raids, the dismissal of all deputies, and free books and tuition.[77]

Despite elaborate plans, both the general strike and the school strike were unsuccessful. This was in part because of the re-arrest of Charles

Guynn on September 4 and Charles Weatherbee the next day. In connection with the two men's roles in urging NMU members to blockade the Spring Canyon highway on August 23 to prevent men not affiliated with the NMU from working, they were charged with riot and criminal syndicalism.[78]

Although the arrest of Weatherbee and Guynn coincided with the release of all but eight prisoners arrested on August 29, the NMU members were infuriated by the jailing of their leaders. Bert Miller, one of the deputies, was beaten up in Helper by NMU sympathizers. Miller was informed that "each and every deputy was to get the same treatment as fast as they could be caught alone."[79] In a meeting held at the Helper Park, Sheriff Bliss was threatened with lynching by a hostile crowd and finally was escorted from the gathering by Mrs. Mike Dragos, a local leader of the Serbian women who was known for her fearlessness during the strike.[80]

With Guynn and Weatherbee in jail and arrest warrants out for Paul Crouch and Harold Huff, the next course of action against the NMU demonstrators was a protest demonstration in the enemy's home territory. Accordingly, a march in Price was planned for Monday, September 11.

In response to what was regarded as an extremely dangerous situation, emergency resolutions were passed by the Carbon County Commission, Price City Council, and Helper City Council to prevent personal injury and property damage. The resolutions prohibited public meetings, restricted all assemblages to no more than three people, placed the sheriff's force under the direction of the county commission, and provided for more sheriff's assistants.

Because of the ordinances, Saturday night dances were cancelled and the state baseball championship, scheduled to be played in Helper on Sunday, September 10, was transferred to Provo.[81]

In defiance of the emergency rules, 400 NMU members met at the Helper Park on Monday, September 11 and were transported in a convoy of cars and trucks to the railroad underpass on the outskirts of Price. Under the leadership of Laurence Mower, chairman of the strike committee, and Ephraim Towne, leader of the NMU local union at National, the men, women, and children formed into ranks four abreast and began marching east down Main Street toward the county courthouse to demonstrate for the release of Guynn and Weatherbee. As Mower and Towne reached the corner of First East and Main Street, they were stopped by L. L. Fryer of the state highway patrol. The strike leaders were requested to march back out of town. When they refused, Fryer grabbed Towne by the arm. At this point the ranks of the NMU marchers broke and led by an unidentified purse-swinging woman they surged forward to protect Towne. Approximately seventy-five deputies and highway patrolmen were on duty and as the marchers started forward, they retaliated with a barrage of tear gas, water from high-pressure hoses, and rifles used as clubs. Main Street resembled a battle zone. A tear-gas bomb shattered the large plate glass window of the Dixon-Taylor-Russell furniture store, and the gas forced evacuation of the store for several hours. The demonstrators quickly scattered and made their way as best they could through vacant lots and back alleys to avoid the deputies. The demonstrators retreated to their parked cars and trucks and returned to Helper but not before the arrest and detainment of four protestors one of whom was Sylvia Crouch, wife of organizer Paul Crouch.[82]

The next morning 100 special deputies moved into Helper at 6:30 a.m. to enforce the countywide civilian martial law and to search for the elusive Crouch. Most of the hotels and rooming houses and many private homes were searched by the deputies in a futile effort to find him. Helper residents were intimidated and frightened by the search tactics of the deputies. Adam Osterich described the way he and his wife were handled by the deputies:

> Upon coming home it was around ten o'clock in the evening, I noticed that they [the deputies] were getting in my yard. . . . After they made a real search in the yard they came and knocked at the door. I told my wife to open the door and ask them for a search warrant, which she did. She got up to open the door. Just the minute that she got to the door Rolla West had his hand already on the hook of the screen . . . and he pulled the screen off first thing. You could see the barrels of the rifles through the door. . . . So then they swarmed into the house. Rolla West ordered me with a gun to sit down in the chair and if I attempt to make a move that he was going to use the gun. My wife asked him if she could turn on the light in the hall for them, so they can see better to go through the house. They told her yes. She stepped over in the hallway to turn on the light and one of the thugs that held the rifle in the hand smacked her on the right side under the ear and floored her down. She was screaming in agony. . . . In the meantime I was getting up to step over and he said if I attempt to move he was going to kill me sure. . . . And I'll tell you that he smelt just like a whiskey barrel.[83]

Fifteen men who participated in the Price riot were arrested in Helper and taken to Price where they were placed in the exhibit building at the county fairgrounds under the watch of an armed guard.[84] Protests against the handling of the NMU and against the arrest of its leaders were registered from many quarters. One of the most effective of these was a meeting held at the First Congregational Church in Salt Lake City on September 15, 1933. Approximately 600 people attended the three and one-half hour meeting. They were addressed by strikers from Carbon County who told of deputies' abuses during the march on Price and of the searches of their homes by deputies looking for Paul Crouch and other local NMU leaders. The coal miners' testimonies were interspersed among the remarks of professional speakers who saw in Carbon County the same oppression that characterized much of America and the world.[85]

In addition to the speeches, a good portion of the meeting was taken up with drafting a series of resolutions protesting the treatment of miners in Carbon County. The resolution called for a delegation to be sent to Carbon County to investigate the conditions. The investigating committee included Brigham H. Roberts, noted historian and Mormon church authority; Warwick C. Lamoreaux, state legislator; Alfred Sorenson, American Civil Liberties Union; Belle Taub, International Labor Defense; Murray E. King, Farm and Labor League of Utah; and Mr. and Mrs. Wilford Owen Woodruff, League for Independent Action. After traveling to Carbon County, the committee held a public meeting attended by several hundred miners, visited court sessions dealing with unlawful acts during the strike, talked with NMU prisoners, and interviewed local officials and citizens. The consensus of the committee members was that local officials were guilty of misconduct—searches had been made without legal search warrants, men and women had been beaten and assaulted, and constitutional rights had been violated. Wilford Owen Woodruff laid much of the blame at the feet of Governor Blood for a policy of "general apathy and persistent procrastination." State Representative Warwick Lamoreaux found that the Carbon County strikers were not anarchists, but miners living and working under feudalistic conditions, having been "held down all their lives by starvation wages." B. H. Roberts, president of the First Council of Seventy of the LDS Church, who died a few days after his visit to Carbon County, noted in his written report that the problem lay with local authorities who favored the UMWA over the NMU.

> It is true that the differences among the miners arose out of controversary between the 2 unions in the field, but it must be added that the county authorities seemingly at least have joined in sympathy with one of these

> rival unions and apparently are administering the law rather partially in their favor, and to the denial of plain constitutional rights to the other faction.
>
> There is an excessive use, and therefore an inexcusable use of power in their favor. Undoubtedly acts of terrorism have been perpetrated upon the least favored parties leading to denial and unwarranted conduct on the part of representatives of the town government that cry out for correction and discontinuance absolutely.[86]

In addition to the protest meeting held in Salt Lake City, other mass meetings were held in Sandy, Midvale, Provo, and American Fork. Written protests and resolutions condemning the lawlessness and the use of force against the Carbon County miners were sent from groups in many Utah communities and from as far away as Minneapolis, Cleveland, New York City, and Worcester, Massachusetts.[87] The nature of these protests suggests that the NMU was part of a state and national network of organizations which, if not communist-directed, at least shared common concerns that Roosevelt's New Deal policies were not sufficient to meet the depression crisis, that the business community was not concerned with the need of the workers, and that a more equitable distribution of the country's wealth was in order.

The International Labor Defense played an important role in protesting the situation in Carbon County and in providing attorneys for the arrested strike leaders. In Utah, the International Labor Defense representative was Belle Taub. She spoke at the September 15 First Congregationalist meeting in Salt Lake City and was elected a member of the investigating committee. In her report, she concluded that the brutalities suffered in Carbon County could only be rivaled "in the terror tales of Fascist torn Germany" and that "in the name of profit—women . . . [and] men are jailed, clubbed, beaten and their blood flows on the streets of Carbon County."[88]

Belle Taub's investigation in Carbon County formed the basis for her pamphlet "The Carbon County Miners' Strike Against Starvation." Issued by the Rocky Mountain District 19 of the International Labor Defense in Denver, Colorado, the pamphlet was sold for a penny and used primarily in an effort to win members for the International Labor Defense. Much of the pamphlet protested the criminal syndicalism charges against Guynn, Crouch, and Weatherbee. Criminal syndicalism was defined as "the doctrine which advocates crime, sabotage . . . violence, or unlawful methods of terrorism as a means of accomplishing industrial or political reform."[89] Belle Taub observed that although Utah had enacted a criminal syndicalism law in 1919, it had never been enforced until the arrest of the three NMU leaders. She warned that if the law were allowed to be used against these three men, every worker stood in danger of being arrested on charges of criminal syndicalism.

Charges against the several hundred miners arrested during the months of August and September were dismissed, and only seven men—Paul Crouch, Charles Guynn, Charles Weatherbee, Laurence Mower, Tony Bonacci, Tom Corak, and Dan Leaner—were held. Belle Taub reported that the International Labor Defense would provide attorneys for the seven men and that committees of workers from surrounding cities, known as "Labor Juries," plus an independent committee of lawyers would attend the preliminary hearings at the Carbon County Courthouse and render their own verdict. She invited Governor Blood to attend the hearings confident that he would become convinced that a grand jury investigation was required.

The hearing for Paul Crouch lasted only two days, October 5 and 6; however, the decision, which was delayed for a week, found sufficient evidence to warrant trial. During the Crouch hearing, Dan Black, a Salt Lake City worker and NMU sympathizer, was taken from the Price City Hall by three men when he refused to obey an order to leave town. Black was driven in a Studebaker to the desert east of Green River, where the three men tied his hands behind his back with his bootlaces, threatened him with death, and beat him unconscious. After regaining consciousness, Black made his way to Green River arriving there at about 4:30 a.m. despite an injured leg, broken teeth, and numerous cuts and bruises. Black caught a ride back to Price with a truck driver. Crouch charged that county authorities were responsible for permitting and perhaps participating in the abduction. Just before the Black kidnapping occurred, a movie photographer from the East reported that his film showing scenes in Price, including a large force of patrolmen, had been confiscated by Price City authorities.[90]

The hearings for Weatherbee and Guynn were postponed until the first week of November, and charges against the four local NMU leaders were dropped. While in jail, Charles Guynn wrote the "Utah Miners Strike Song," an eight-verse ballad sung to the tune of "Casey Jones."[91]

The charges of criminal syndicalism against the three were dropped. Paul Crouch and Charles Guynn were tried on charges of riot; however, because of ill health, Charles Weatherbee was not brought to trial. Both

Guynn and Crouch were found guilty by Carbon County juries. In an appeal to the Utah Supreme Court, Charles Guynn succeeded in having the conviction set aside and the case remanded to the Carbon County District Court for a new trial. Guynn's attorney, H. G. Metos, argued before the Utah Supreme Court that R. L. Loveless, foreman of the jury which convicted Guynn, had served as a deputy sheriff during the Spring Canyon raid and on numerous occasions had described the NMU and its leaders as a "bunch of reds and ought to be run out of the country."[92]

The trials effectively weakened the NMU in Carbon County and allowed the UMWA to complete its countywide organization without any great interference. Nevertheless, the NMU still tried to maintain a foothold. NMU miners struck the Mutual mine for a third time on October 25 when Superintendent Sutch announced that the bimonthly pay schedule was being postponed. The strike lasted only one day, and miners returned to work when Sutch agreed to meet the October 25 payroll.[93]

This was the last success for the NMU. In April 1934, the Helper City Council prohibited the holding of an International Labor Defense Conference in Helper. Under the leadership of Mrs. Charles Guynn, the conference was to discuss ways of freeing Paul Crouch and Charles Guynn.[94]

A May 1 parade scheduled to march through Helper attracted only forty participants and was eventually abandoned. A demonstration was held in Helper to protest the town's resolution prohibiting the International Labor Defense Conference. Circulars also demanded "a decent relief station in Helper, adequate cash relief for all unemployed, and the passing of an uninsurance bill." The only disturbance came when Mrs. Margaret Nemanich approached Helper Mayor E. F. Gianotti and offered to let him "take a smell of relief ham." Before the demonstration began, five people were arrested, including Mrs. Nemanich and four others for distributing "communist literature." Sheriff Bliss had anticipated serious trouble and had assembled 100 special deputies to deal with the scheduled parade and demonstration.[95]

A final incident occurred on June 17, 1934, when a mass meeting held in the Helper City park was broken up by Helper policemen and firemen under the direction of Mayor Gianotti. The policemen attacked the assembly with guns and clubs while the firemen "drenched the speakers and audience with dirty canal water."[96] Mayor Gianotti was the subject of resolutions protesting his violent and irrational conduct, especially in light of his profession as a medical doctor.

Despite the hostile relations between the UMWA and the NMU, reconciliation did come. In its September 29, 1933, issue, the *Helper Journal* reported under the headlines, "After the Storm The Calm":

> During the past week it appears that much of the apparent ill-will and hard feeling which has for several weeks existed between various factions, for reasons which may or may not have been justified, is falling by the wayside, and that conditions are again bearing a resemblance to normalcy.
>
> . . . the attitude which now is being adopted by members of the National Miners Union and encouraged by its leaders—a spirit of friendship and good will to miners of other unions with a view to solidifying the ranks of the workers.[97]

However, it was not until more than a year later that the last vestiges of the NMU were eliminated from Carbon County when the UMWA finally succeeded in organizing the Maple Creek mine. On December 6, 1934, Frank Bonacci described the end of the NMU in a lengthy communication to John L. Lewis.

> On Sunday December 2nd I organized the Maple Creek Mine and I have taken in 50 miners former members of the "National Miners Union including their leader Charles Guinn." This man for the past 18 months has been the leader of the National Miners Union and the Comunist Party, and played Hell both in Utah and New Mexico, the Maple Creek Mine was the stronghold of the so-called National Miners Union, some time ago I had started an organization at this mine with 14 members but we met opposition from this other organization and we failed to make any headway, the company then took a hand and discharged three of our local officers result was that a strike was called the mine was picketed and after two days the company agreed to re-instate our three members, our men being in minority were abused and they found jobs elsewhere and I had to take the Charter to my home.
>
> A week ago they called a strike because they had three pay days due them, and I took advantage of that and again made another attempt to organize them by convincing them that the United Mines Workers of America was the only organization that could give them protection and as result that we now have 100% organization at that mine, I am now waiting for Dist Pres, John M. Ross who is due here next week to contact this company and ask a working agreement. I had to take in Charles Guinn in order to organize that mine in which he is an employee, but I have done so with the understanding that he can not hold any local office and his membership is subject to your approval, in his defence

> Guinn claims that at the time he became an officer of the dual organization he was not a member of our union therefore he is eligible to join. . . . I asked him if he was ready and willing to forfeit his membership in the dual organization and his answer was, *Yes I do.*
>
> Of course Charles Guinn if permitted in our Union his activties must be watched, he fooled many Utah miners for so many months, at one time he had about 500 members, but with the signing of the agreements the majority joined our Union few of the most active members of that organization refused to join and . . . [retreated] in the smaller mines, but since we have now also organized the small mines his purpose for joining our Union may be as President Green warns "Boring From Within."
>
> The organizing of the Maple Creek Mine their stronghold if we can secure a contract and the check off at that mine means "The Funeral of the National Miners Union in Utah." We are now organized 100% and working under contract with full protection of the United Mine Workers of America, only one mine remains not working under contract, except the Wagon and Truck mines not yet all organized, the largest of them employing 25 miners, three others 15 each and the others from 3 to 6 men in each mine, less than 200 miners are employed in the non-Union Wagon mine in Utah.[98]

Lewis replied that he concurred in allowing Guynn to join the UMWA and expressed the hope that Guynn would "in the future comply with the laws of the United Mine Workers of America and observe its policies."[99] Apparently Guynn did conform to UMWA laws and policy. Rolla West reported that three or four years after the strike, Frank Bonacci informed him that Guynn had been hired as an organizer for the UMWA by John L. Lewis. Bonacci reportedly explained, "Old John needed agitators bad, and Guynn was good at it."[100]

Although the UMWA was welcomed with open arms when it offered a more conservative alternative to the NMU, a few coal operators tried to circumvent UMWA organization of the mines. Some operators and managers saw in the two competing unions the opportunity to create "a situation making it possible to play one union against the other until finally eliminating both unions."[101] On September 7, 1933, Nicholas Fontecchio wrote Philip Murray, UMWA vice president, that "the United States Fuel Company and the Columbia Fuel Company mines are still nursing the National Miners Union hoping that by the split they will be able to avoid dealing with the . . . [UMWA]."[102] Earlier A.T. Fotland, secretary of the Standardville local of the UMWA, wrote to William Knerr, chairman of the Utah Industrial Commission, that the Spring Canyon Coal Company was discriminating against UMWA members in an apparent effort to "keep the men split up." In his letter, Fotland also disclosed anti-UMWA sentiment by the United States Fuel Company. "Last night we organized 41 men from Hiawatha into the U.M.W.A. and that was done against the desires of the U.S. Fuel Company."[103] The resistance of the United States Fuel Company was sufficient to warrant Fotland suggesting that company officials be arrested and prosecuted as an example to others who were resisting the unionization drive.

Perhaps it was Columbia Fuel Company officials whom UMWA officers wanted to impress with the federal support of their organizing activities. At the Columbia mine, located in the eastern part of the Carbon County coal fields, company officials steadfastly refused to recognize the UMWA. During the fall of 1933, twenty men were dismissed because of the active part they took in organizing a local union.[104]

Unlike the NMU, which had little faith in the New Deal programs and sought to win success through strikes during the early weeks of the New Deal, the UMWA chose to establish its position as best it could without resorting to strikes and to wait and see what kind of coal-mining code would be adopted under the National Industrial Recovery Program. Title I of the National Industrial Recovery Act sought to end cutthroat competition among businesses and within industries, to raise prices to a profitable level by limiting production to actual needs, and to guarantee the American worker a reasonable workweek and a living wage. This was to be accomplished through the adoption of codes for all branches of business and industry by committees representing management, labor, and the public.

On July 6, 1933, fifty-five coal operators and officials from New Mexico, Colorado, Wyoming, Montana, Washington, and Utah met at the Newhouse Hotel in Salt Lake City to draft a uniform code for the western coal industry. All delegates agreed that unfair competition must be eliminated before the coal industry could be restored to a profitable footing.

In addition to the fair-competition provisions, the conference considered three items of paramount importance to the coal miners: a daily maximum number of hours of labor, a weekly maximum number of hours of labor, and a minimum wage per hour for each class of employment. The bituminous coal code adopted in Washington, D.C., on September 16, 1933, established $5.44 per day or sixty-eight cents an hour as the basic rate of pay for skilled workers. Unskilled laborers were to receive $4.48 per day or fifty-six cents an hour. Maximum hours of work were set at eight hours a day and forty hours a week. In addition, the code contained the following conditions of employment: employees were guaranteed the right to organize and bargain collectively through their representatives; no employee would be required to join a company union as a condition of employment; miners were to be paid for all coal mined on the basis of a 2,000 or 2,240 pound ton; miners were to have a checkweighman of their own choosing, and the right to check the accuracy and fairness of weighing methods used by their representatives; wages were to be paid semimonthly; employees were not to be required as a condition of employment to live in company-owned housing; employees would not be required to trade at a company-owned store; no one under seventeen years of age would be permitted to work inside a mine; and no persons under the age of sixteen would be employed for work outside a mine.[105]

Three weeks after the National Bituminous Code was signed, representatives of UMWA District 22 met with officials of the Utah Coal Producers and Operators Association at the Newhouse Hotel in Salt Lake City to develop an agreement covering the Utah miners. After a month-long conference, a joint contract, which conformed to the provisions of the National Bituminous Code, was signed on November 8, 1933. The *Wyoming Labor Journal* celebrated the occasion by noting that after more than forty years of nonunion operation, the Utah miners and operators had adopted the principle of collective bargaining.[106] Upon notification that the Utah agreement had been signed, UMWA Vice President Phillip Murray wrote to Nicholas Fontecchio commending him for his success and observing, "This agreement is a very substantial contribution towards stabilization of the bituminous coal industry, through the medium of uniformity of labor costs in the Utah territory."[107]

The only major coal company that failed to sign the Utah agreement was the Columbia Steel Company. Its reluctance was in line with the steel industry's nationwide policy of "avoidance and eradication of unionism among its workers."[108] The steel companies believed that the unionization of their coal miners, who produced about 8 percent of the national tonnage for their own consumption, would lead quickly to unionization of their mills. Steel officials maintained that conditions in the captive mines (mines operated by steel companies for their own use rather than coal production for a commercial market) were much better than in the rest of the coal industry—miners were paid good wages and working and living conditions were excellent.

John L. Lewis countered that there was little difference between the commercial and captive mines and encouraged his organizers to bring the miners of the captive mines into the UMWA. The steel companies responded by organizing company unions and dismissing miners active in the UMWA.

Lewis waited until the completion of the Bituminous Coal Code on September 16, 1933, and the Appalachian Agreement for anthracite coal five days later before demanding that the steel industry accept identical terms for the captive mines. The steel companies offered their miners the wages, hours, and other provisions of the commercial agreement but denied recognition of the union and refused to institute the checkoff system to collect union dues. At this point, Franklin D. Roosevelt entered the controversy by stating that he felt the checkoff should be included in any agreement, provided the men asked that money be collected from their pay. Under strong pressure from Roosevelt, an agreement was reached on October 30 whereby miners in the captive mines would be awarded the same wages, hours, and conditions as those of the Appalachian Agreement; miners who had been dismissed for union activities would be recalled without prejudice; the checkoff system would be implemented; and the National Labor Board would hold elections in which the miners would choose their representatives for collective bargaining. It did allow the UMWA to represent those mines in which it had organized a majority of the workers, thereby providing a substantial advantage for further UMWA unionization of the steel industry.[109]

The situation at the Columbia mine in Utah reflected on a local scale the same circumstances and solution that characterized the national scene. UMWA Local No. 6089 was established at the Columbia mine sometime before mid-August 1933; however, the company officials, under orders from the San Francisco office, refused to recognize the union and sought to dissipate its strength at every opportunity. On August 18, 1933, the local secretary, Orvel Peterson, reported that "Stool Pigeons" had paid the fifty-cent initiation fee to join the UMWA but were "setting on the fence so if the company wins out they can fall their way and if we win out they can fall our way."[110]

In November James Morgan reported that he had attended a meeting of the Columbia miners to discuss the posting of a notice announcing the work force would be reduced. "Our members were very much worked up over the matter, believing that it was a move that would largely get rid of the active men in the organization."[111] The men who were dismissed remained in camp with financial support from all local unions of District 22 sustaining them.[112]

Despite the willingness of the Columbia miners to strike, John L. Lewis advised them, through District Vice President George Lambert, to remain at work until the National Labor Board issued its decision on the captive mine situation.[113] By February 23, 1934, the patience of the Columbia miners ran out and they declared a "holiday" to demand the reinstatement of men laid off and UMWA recognition. John L. Lewis quickly recommended that the men return to work, fearful that the wildcat strike would jeopardize the decision of the National Labor Board.[114]

The same day that John L. Lewis urged the Columbia miners to return to work, February 24, 1934, George Creel, a member of the Regional Labor Board in San Francisco, delegated to Allen T. Sanford, Utah State director of the National Emergency Council, authority to deal with the situation at Columbia.[115] F. E. Morris of the NRA Labor Compliance staff was sent to Columbia. He secured a promise from the company that those men who had been discharged would be reinstated by March 12. The company also agreed to allow its miners membership in any union they selected and to institute a checkoff system.[116] The settlement was based on the same arrangement that had been followed in other captive mines. On March 14 three members of the divisional labor board —Chairman T. S. Hogan of Denver, James Morgan of Cheyenne, and D. D. Muir of Salt Lake City—arrived in Columbia to supervise the voting on union affiliation. Of the 113 votes cast, 105 favored UMWA representation during collective bargaining. Two days later, Friday March 16, District 22 officials and officers of Columbia Local No. 6089 met with W. P. Phipps, general manager of operations for the Columbia Steel Company, and Walter Shelton, attorney for the company. An agreement that incorporated the provisions of the Utah Coal Code was signed the following day.

By the end of 1934, the UMWA was firmly established in the Utah coal fields. The coal companies, which had opposed unionism since the coal fields were opened, offered little resistance to the UMWA organizational drive and agreed to a coal code addressing the long-standing grievances expressed in earlier labor disputes. The Columbia mine, Utah's only captive mine and the UMWA's most stubborn opponent among the coal companies, was forced by the overwhelming vote of its employees to recognize the UMWA. By December 1934, the last vestiges of the NMU were absorbed into the UMWA as Charles Guynn and fifty former members of the NMU at the Maple Creek mine joined the UMWA. The success of the UMWA in 1933 and 1934 marked the end of a long chapter in the history of both Utah and the nation. No longer would the struggle to establish a miners' union in the Utah coal fields characterize labor-management disputes as it had since 1901.

Chapter XI

The Union Legacy

With the organization of the American coal miners virtually completed, John L. Lewis announced in 1935 his more ambitious goal to "organize the unorganized."[1] The vehicle for this expanded effort was to be the Congress of Industrial Organization (CIO), established in 1935 under the principal leadership of John L. Lewis, who was elected the organization's first president. The CIO set its sights high, and by 1938 it organized the important auto and steel industries.

Paralleling the national aims of John L. Lewis through the CIO, in June 1936, union leaders in Carbon County organized the Carbon County Central Labor Council, composed of the local UMWA unions, auto mechanics, bricklayers and stonemasons, bartenders, and store clerks. George B. Harding of Price served as president; Robert B. Henderson of Kenilworth as vice president; and Frank Bonacci as secretary-treasurer. The first objective of the local council was to "organize all unorganized workmen."[2]

Frank Bonacci remained with the UMWA until August 1937 when John L. Lewis appointed him a field organizer for the CIO for $225 a month. Bonacci continued with the CIO and elected to stay with the organization when the UMWA broke away in 1942. In May 1942, Bonacci was appointed regional director for the CIO in Colorado. Later, his area of responsibility was expanded to include Utah and New Mexico. After a reorganization in 1943, Bonacci served as regional director for Utah and Nevada until October 1951 when he suffered a heart attack and retired from the CIO on July 1, 1952.

In 1936, Bonacci was elected to the Utah House of Representatives from Carbon County and served six consecutive terms until 1952 when his health precluded his seeking reelection. As a legislator, Bonacci worked strenuously for worker protection benefits for workers and the aged. His concern with education resulted in the establishment of Carbon College, now the College of Eastern Utah, in 1938. Following his heart attack in 1951, Frank Bonacci endured poor health until January 20, 1954, when he died in the Price Hospital.[3] No other man better symbolized the struggle for unionism in the eastern Utah coal fields. In death, the revered union leader and esteemed political servant symbolized the labor union's respectability and power in Carbon County.

The experience of 1933 has left Carbon County with a unique legacy. Of nonagricultural workers, it has the highest percentage of union members of any county in the state.[4] When Roosevelt upheld the right of labor to organize he created a stronghold of Democratic support that continued through the 1980 election when Carbon County was the only one of Utah's twenty-nine counties to support the Democratic presidential candidate, Jimmy Carter; the other twenty-eight voted overwhelmingly for Republican Ronald Reagan. However, the 52 percent that Carter received in 1980 indicated a significant erosion from the support rendered Roosevelt. In the three presidential elections after the organization of the Utah coal fields, Roosevelt's popularity was evident: 79 percent of the total county vote in 1936, 71 percent in 1940, and 70 percent in 1944. In some of the coal-mining precincts the votes for Roosevelt's opponent could be counted on two hands. Percentages in favor of Roosevelt reached 94 percent in Clear Creek in 1936 and just under 90 percent in Standardville, Rains, Sunnyside, and Consumers. Even in 1940, when UMWA President John L. Lewis announced his support for Wendell Wilkie because he thought Roosevelt had deserted union workers, the Carbon County coal-mining precincts of Scofield, Clear Creek, Royal, Kenilworth, Rains, Columbia, and Sweet registered more than 80 percent support for Roosevelt. All the mining precincts in Carbon County voted 1,849 to 537, or 77.5 percent in favor of the third-time Democratic candidate.

After 1934, the history of labor relation issues was tied primarily to national issues with regular focus on contract negotiations. When these negotiations were not completed on time, Utah miners walked out until ordered by John L. Lewis to return to work. However, the solidarity of the Utah union men was not tested during the remainder of the 1930s. When walkouts were called beginning on April 1, 1941, the wartime demand for coal made quick settlement a necessity. The rampant inflation that broke loose during World War II led John L. Lewis to demand a wage increase for his miners. The negotiations dragged on (with several

Utah walkouts) until a four-day work stoppage in November 1943; at that time, a new contract was negotiated, raising the miners' base pay from $8.12 to $8.50 a day.[5] During July 1943, coal production reached only 50 percent of the 1942 level.[6] However, miners blamed the difficulties on the federal government, specifically on the National War Labor Board, which tried to keep wages at the 1941 level.

With the coal mines being worked six days a week and with mine operators calling for an additional 1,400 miners to supplement the 3,200 already employed, the coal companies were anxious to keep production at the highest possible level. The coal companies recognized that any increase in the cost of coal would have to be absorbed by the government and would not diminish their own profits.

Unsanctioned walkouts also occurred in Utah when miners failed to receive back pay within a reasonable time, when they felt their bosses mistreated them, and when the company raised hospital fees from $3.50 to $5.00 a month in the newly opened Kaiser mine in Sunnyside and then refused to rent company buildings to the union when it proposed to operate its own medical and hospitalization program. When these wildcat strikes occurred, district and national union officials asked the men to return to work.[7]

Utah miners demanded what they believed they were entitled to and, along with their national counterparts, were willing to reinforce demands by work stoppages. Because of the national emergency, they incurred considerable criticism: the coal miners were vital to the national war effort, and, in the final analysis, their sacrifice matched those who served in the armed forces. Between January 1941 and December 1945, a total of 140 miners died in Utah's coal mines—55 more than the 85 Carbon County residents who died in war service.[8]

Mine disasters continued during the war. Seven men were killed at Kenilworth on March 14, 1945, when a pillar of coal collapsed severing an electrical cable, which short-circuited on a metal pipe, creating an electrical arc that ignited coal dust and methane gas. Twelve men were burned, seven died. Two months later, in May 1945, an explosion of gas in the Sunnyside mine left twenty-three miners dead.

The prosperity of World War II continued through 1947, when nearly 7.5 million tons of coal were mined by over 4,000 miners, most of whom were members of the UMWA. After 1948, a steady decline in the coal industry began, which continued for more than two decades until the end of the 1960s. By then, there were fewer than 1,500 working coal miners producing just over 4 million tons of coal.

The energy boom of the 1970s was a dramatic and, in many ways, painful reversal of the quiet years after World War II. The magnitude of the 1970s boom was overwhelming in its impact on Carbon and Emery counties. The peaceful Mormon communities grew at an alarming rate with an influx of new miners whose different lifestyles and backgrounds contrasted and often conflicted with traditional Mormon and old-world traditions and practices which were woven together in a unique fabric in the eastern Utah area. Pastures in towns like Huntington and Castle Dale, where horses, sheep, and cattle had grazed for nearly 100 years, became instant trailer courts. The road on which teams and tractors had pulled hay wagons or on which cattle and sheep had made their way from summer ranges in the mountains to winter ranges on the deserts were filled by roaring coal trucks and by four-wheel-drive pick-up trucks (which cost young miners the equivalent of what many earlier miners had earned during the entire decade of the prosperous 1940s). Earnings of the new miners for one day often exceeded the amount their predecessors of the 1940s had earned in two weeks, working six days a week. The new boom of the 1970s had no comparison with earlier periods in Utah's coal-mining history and sorely challenged community leaders to meet the expanded demands for housing, schools, roads, stores, recreation, and other needs.

For the UMWA, the boom tested local union leaders in ways reminiscent of the pre–1933 period. Until the 1970s, union membership encompassed all but a small group of men who worked a few family and wagon mines in the area. However, as new mines were opened, most of the operators sought to provide pay plans and insurance programs to rival the benefits provided the union miners. As a consequence, nonunion mines made significant inroads into the previously all-union coal fields of eastern Utah.

Antagonisms exist between union and nonunion miners. Nonunion miners maintain that the union has outlived its usefulness, has become too repressive of the miners' rights, and has become too prone to scandals and misconduct by its leaders. Union miners reiterate that the nonunion miners enjoy their benefits because of the union threat to management; and when the cycle of hard times returns to the area, the nonunion miners will be left with no protection in their negotiations with the coal operators.

The UMWA continues the struggle for better working conditions, improved safety conditions, higher wages, expanded benefits, a more liberal pension program, and organization of the nonunion mines. Yet the benefits and earnings of the miners are among the best and highest in the state. Today, fifty years after union recognition, Utah's union miners face the challenge of efficient and effective production, especially as their work is matched against that of the nonunion miners. Furthermore, the economic recession of the 1980s has hit Utah's coal fields with unexpected and severe impact. Work is being curtailed and men, who only a few years ago were promised a bright future as coal miners, have been forced into the ranks of the unemployed. Such challenges will continue as long as the coal industry remains an essential pillar of Utah's and America's economy.

Chapter XII

Conclusion

Speaking to the delegates of the Western Federation of Miners at their Sixteenth Annual Convention held in Denver, Colorado, Roderick MacKenzie remarked "the average native of Utah is . . . the hardest biped to unionize in all America."[1] Although MacKenzie's perception was accurate, especially for 1908, it did not consider the complex issues that essentially validated his judgment. As suggested in the previous pages, the long and protracted struggle for union recognition in the Utah coal fields was the product of an intricate combination of factors whose strength ebbed and flowed in response to economic, political, and other issues. Nevertheless, Utah's experience differed from neighboring Colorado's and Wyoming's and from other coal-producing regions of the country. A review of the forces and elements that shaped labor relations in the Utah coal fields illustrates this complexity and allows for some assessment of their significance in this story.

Until 1906, the eastern Utah coal industry was completely under the control of the D&RGW through its subsidiaries, the Pleasant Valley Coal Company and the Utah Fuel Company. The land fraud cases against Utah Fuel and a continuing coal shortage in Salt Lake City led to the opening of the first independent mine at Kenilworth in 1906. The development at Kenilworth was soon followed by new mines at Hiawatha and Mohrland. World War I stimulated the growth of recently opened mines in Spring Canyon; and by the early 1920s, the last area, known as Gordon Creek, was opened. Despite the number of different coal companies operating in the Utah coal fields, their stand against organized labor was united. They refused to recognize any labor organization formed by the miners; and during strikes, company officials steadfastly refused to talk with officers and organizers for the miners' union. Although 1933 brought great change and most of the coal companies could see that UMWA recognition would be a necessity, there were still companies like United States Fuel and Columbia Steel that sought to follow the traditional course.

The Utah coal operators successfully avoided the situation that George F. Baer, head of the Philadelphia and Reading Railroad, confronted when he was forced to submit to negotiations with the UMWA during the famous 1902 anthracite strike. The Utah coal operators were much more conservative than the general Utah business community, which wholeheartedly supported the right-to-work doctrine. Although the right-to-work philosophy provided that both union and nonunion men could work side by side, the attempts of the UMWA in the 1903 and 1922 strikes to secure company recognition (thereby permitting the union men to return to work without any discrimination) were unsuccessful.

During 1918 and 1919, some companies, notably Independent Coal and Coke at Kenilworth, were willing to allow the existence of a union organization as long as the local union did not demand the right of collective bargaining and did not hold its meetings on company property. However, this small sign of tolerance was short-lived; and as the coal industry began its descent into the depressed 1920s, the local unions were quickly forced to conduct business in secret.

Well-tried tools and methods were employed to repress the interests in unionism: company spies were used to keep tabs on union leaders; known union men were discharged; blacklists were circulated among the different companies; and yellow-dog contracts were used in some cases. In addition, the coal companies made a conscious effort to maintain a balance of diverse ethnic groups to deter solidarity. Each company town had its athletic teams and amusement hall, which were devices to channel potential grievances into an unrelated arena.

During much of the period covered in this study, the coal companies controlled local politics. Their control was successfully, but briefly, challenged by the Progressive party during the 1912 election.

Although coal companies argued that union recognition would bring economic hardship, it appears that the companies were opposed to the labor unions more on ideological than economic grounds. In Utah, the miners were paid wages equal to those of the union mines in Wyoming and Colorado. Despite the recognition of the local unions in Wyoming and Colorado, the miners there still voiced the same grievances regarding the company store and inaccurate weighing of coal as did their counterparts in the unorganized fields of Utah. In this manner, recognition or nonrecognition of the union made little difference before 1933, except as a moral victory in forcing the coal companies to negotiate with union officials. Had the Utah miners not been intimidated by the coal companies, the UMWA would have become a permanent institution in Utah long before 1933.

The work force of coal miners was usually divided into two groups—the American miners and immigrants from English-speaking countries who usually became American citizens in a short time and the non-English-speaking immigrant miners, the majority of whom were from Italy, Greece, the Balkan countries, and Japan.

Although there were many English-speaking miners who were leaders during the strikes and were instrumental in the efforts to establish organized labor in Utah, a significant number refused to support the unionization attempts.

The chance for upward mobility in the coal industry was much greater for the English-speaking miners. Anxious to secure the more stable positions of company men, those with such ambitions realized the harm that union affiliation could do to their chances for advancement. The situation in the coal fields was somewhat like that in the steel industry:

> "Each man is out for himself," interviewers were told repeatedly. "The men here have lost any spirit of working together. . . . They look out for their own interest." One worker observed of the higher paid men during the 1919 organizing campaign: "Haw, they won't join no union; they're all after every other feller's job." To encourage this spirit, managers held out unlimited prospects for advancement. . . . Insecure and ambitious, they came to identify their fortunes with the good will of their particular boss, and they behaved accordingly.[2]

The English-speaking men occupying the supervisory positions favored those like them. These men were assigned the best working places, were given jobs in preference to the immigrant miner, were provided with company houses before the immigrant, and during a slack period, were given available work first. Many feared this favorable arrangement would end if the coal companies were forced to negotiate with a labor union.

In addition, the coal company officials appealed to the pride of the American miners. During the strikes, it was stated that a miner who supported the strike was not man enough to earn what he needed on his own in competition with others but rather was obliged to seek the help of a labor organization, not because of any injustice by the company, but because of his own weakness and inability.

The animosity between the English-speaking and foreign-speaking miners proved to be a great hindrance to the successful unionization of the coal fields. The immigrant was looked upon with suspicion, and at times, fear. He was regarded as inferior by the American miner, who consistently criticized his unwillingness to give up native customs and languages and to become "Americanized." Martin Cahill, president of District 22, noted the difficulty in a letter to John L. Lewis regarding 1929 organization efforts in Utah.

> In our opinion, if the Native or American, speaking Miner would take an active part or the lead, the Foreign speaking Miner would follow. There is quite a bitter feeling existing between the American and Foreign Miner, and on account of this feeling it is doubtful . . . if anything could be done alone the lines of orgnizatin unless the American Miner would take the lead.[3]

Cahill's observation was correct. A divided labor force offered little hope of a successful organization. Not until 1933, when the majority of American miners joined willingly with the immigrant miners, did a labor union firmly establish itself in the Utah coal fields.

Because of the demand for coal and coke and the shortage of laborers and miners, it was inevitable that immigrants who were part of the great influx of the late nineteenth and early twentieth centuries would be recruited for the Utah coal mines. Being the greater number, the foreign-born miners played a crucial role in the history of organized labor in Utah. They came from a wide variety of backgrounds. Some, especially those from the industrial centers of northern Italy, were acquainted with labor unions through their experiences in the old country. Some brought socialistic and even anarchistic sentiments to Utah. Others, from the small farming villages of southern Italy and Greece, were not at all familiar with the goals of organized labor until after they began work in Utah. A number received their union indoctrination in other parts of the country—Pennsylvania, Colorado, Montana—or, as with Frank Bonacci, in Canada. In short, there was no common labor union experience for Utah's immigrant miners before they reached the state.

Early labor historians decided that the immigrant was a detriment to labor organization. John R. Commons found that recent immigrants were

> unfamiliar with the traditions and customs of labor organization, unaccustomed to the rules and control which it imposes, incapable of unionizing through their ignorance of the language, and, moreover, forced by their poverty to work for low wages, and, by their lack of friends to work with docility and desperate energy for him who first gives them a job. . . . The constant influx of immigrants into a trade nullifies . . . the educational work of organization almost as rapidly as it is conducted.[4]

Later students have argued that immigration was in large measure responsible for adopting a nonradical business union position. Although native-born workers reacted with radical means to the upheaval caused by rapid industrialization and modernization from 1880 to 1920, immigrant workers were much more conservative, interested only in the immediate economic benefits that characterized the philosophy of business unionism.[5] Others have placed immigrants at the center of the American labor movement serving to aid unionism in one instance and hinder its development in another. Robert H. Zieger writes:

> Immigrants crucially influenced the course of American labor history. Sometimes as strikebreakers, sometimes as militant supporters of collective action, they both frustrated and furthered the rise of unions. Deeply imbued with ethnic consciousness, they often proved indifferent to the union or radical appeal. On other occasions they drew strength from their neighbors, churches, and traditions, and courageously resisted exploitation. They were heroes and apostates, schemers and victims. Their story is an historical story, taking place in specific cities, workplaces, and neighborhoods. It is the story of subtle adjustments, specific responses rooted in specific cultural traditions.[6]

In Utah, instead of breaking the trend of American miners to more radical action, the immigrants were a catalyst for increased awareness and for change of the exploitive policies of the coal companies. Their intentions were more than economic, and their actions were often in response to a strongly felt indignation against injustices rather than a desire solely for economic gain.

The experience of Utah's immigrant miners closely paralleled that of the immigrant workers in the steel industry. In his book *Steelworkers in America: The Nonunion Era*, David Brody writes:

> The mass of peasant workmen were, as they appeared, unpromising union material. Nothing in their agriculturalist past prepared them for collective activity in industrial America. Uncommitted to a permanent

> life in the mills, moreover the immigrant did not find the arguments for joining a union persuasive. He knew that the unions spelled dismissal or at least a prolonged strike . . . What was a dubious small increase compared to several months of idleness, when each day out of the mill added two days to his exile. There was no interest in trade unionism among recent immigrants in steel towns.[7]

However, the lack of interest in trade unionism lasted only a short time, and the immigrant workers proved dedicated members and strikers when they once embraced a labor organization. Brody continues his description of the immigrant:

> His ability as a striker sprang from his European background. The peasant looked on himself primarily as a member of a family and a village. He could not conceive of differing with group decisions, unanimously reached. Moreover, nothing mattered more than communal approval. Immigration weakened, but did not destroy, group consciousness. When, therefore, immigrant workers were driven to strike, they were likely to persist until the general sentiment turned. They were effective strikers because they were peasants.[8]

The pattern outlined by Brody held true for the immigrant miners in Utah. The first large group of immigrant coal miners were from Finland. They entered the mines of Winter Quarters and Clear Creek during the 1880s. By 1901, they had become sufficiently disillusioned with conditions that they became an important element in the strike of that year. The next group of immigrants, the Italians, arrived in the 1890s and were concentrated at the newer Castle Gate and Sunnyside mines. Though perhaps sympathetic with the Winter Quarters and Clear Creek strikers in 1901, the Italians refused to join with them. However, less than three years later, during the 1903–4 strike, the Italians were the most committed participants in that labor dispute. The places of the Italian strikers were filled by Greek immigrants of peasant background who only wished to earn sufficient money to return to their homeland and a better economic condition. Yet shortly after their arrival, they began to protest the abuses of the padrone system and by 1911 were willing to strike because of cheating by the company at the weighing scales. When the UMWA reentry in the Utah coal fields was climaxed by the strike of 1922, the Greek miners were the most loyal strikers. Eleven years later during the National Miners Union strike in 1933, it was the Slavic miners who formed the nucleus of the strike.

The fraternal organizations and community life of the immigrants helped develop the solidarity displayed during the various strikes. Nevertheless, the lack of unity between immigrant and native miners served the interests of coal company officials who hoped to discourage unionization by maintaining a balance of different nationalities in each camp. Blame for the lack of cooperation cannot be placed on the English-speaking miners alone. In describing the relationship between native Mormons and immigrant Greeks, Helen Zeese Papanikolas astutely observes:

> To the Mormons the Greeks were interlopers among the "white" population; they were clannish, would not marry outsiders, and thought they were an exceptional people with the only true religion on earth.
>
> The Greeks on the other hand, thought the Mormons, a high percentage of them of Nordic strains, dull people "without salt." They called them "whiteheaded" and "inhospitable." "They wouldn't give you a glass of water if you were choking to death." Greeks avoided certain restaurants in small towns; they had either been ignored or rudely served. The Mormons were also clannish, would not marry outsiders, and thought they were an exceptional people with the only true religion on earth.[9]

Until the 1880s, the Mormon church felt relatively little concern for organized labor. Church leaders were generally from rural areas where unions were nonexistent. After the church was transplanted to Utah, several small trade unions developed, but because of their conservatism and the harmony of their actions with general church objectives, there was no opposition from church leaders. The economic program of cooperativism had as its ultimate goal the improvement of life for church members while remaining free from dependency on non-Mormon operated industries and businesses. In large measure, these were also the economic goals of organized labor: the improvement of living conditions while insuring that the laborer was not in complete subjugation to his employer. Historically, both movements sprang from the same reformist ideas of the 1830s; however, within fifty years, a wide gulf was glaringly evident.

The completion of the transcontinental railroad in 1869 brought great change to Utah's economy. The railroad made possible the development of Utah's mining industry, and this, in turn, greatly stimulated other aspects of economic life. The railroad also led to a great influx of non-Mormons, and the resulting conflict between Mormon and non-Mormon was one of the most intense struggles in American history. One aspect of the larger confrontation involved the activities of the Knights of Labor. Active in Utah and the Intermountain West during the mid–1880s, the organization was led by non-Mormons, appeared to advocate violence in excluding Chinese laborers, proposed cooperatives somewhat similar to those developed by the church, and was known as a secret organization. All of these characteristics were unacceptable to church authorities, and their experience with the Knights of Labor initiated the development of an extremely conservative policy towards organized labor. Kenneth Davies writes that by the early 1900s:

> There appears to have been a basic shift in the attitude of the Church leaders towards unions. Previously, they had on occasion defended unions, even in some of their strikes where exploitation of workers was strongly evident. Less and less was there to be seen positive support of unions and their leaders, and more and more were they to be derided.[10]

During the first decade of the twentieth century, church leaders began to articulate their opposition to the closed shop by denouncing it as contrary to the divine right of free agency. Church authorities argued that joining a labor organization as a prerequisite for employment violated the God-given right of free choice. On the other hand, organized labor was convinced that without the closed shop, unions were attempting to build a house on a foundation of sand.

It has been suggested that the use of the free-agency issue by church leaders was only a screen for anti-union sentiment of the conservative business community from which the church authorities had come. The attitude of the church toward the unionization attempts in the coal fields sheds some light on this contention; despite the importance of the church-union conflict over the closed-shop principle in Utah labor history, it was of little direct importance in the Utah coal fields during the period of this study. Before 1933, organized labor was never in a position to demand the closed shop in the coal industry. The UMWA failed to secure the right for coal miners to join the union. Nevertheless, the Mormon church did serve to retard the efforts of the UMWA in the Utah coal fields. In 1883, it was Abraham O. Smoot, president of Utah Stake, who persuaded miners at Winter Quarters to return to work. During 1901, local church leaders issued statements denouncing the strike, and newspapers explained that it was the "non-Mormon" miners who were out on strike. In 1903–4, the conflict between the church and the UMWA reached a boiling point when Angus Cannon, Salt Lake Stake president, announced that jobs were available in the coal fields despite the fact that a strike was in full swing. The opposition by the church was an important factor in the lack of organizing activity by the UMWA until 1918. John McLennan reported to William Green, "The Mormon Church, also heretofore bitterly opposed to the labor movement, has changed its attitude and will no more oppose us."[11]

Despite McLennan's optimism (and from available evidence, the church did not take a direct stand against the UMWA), the union met with much greater success among the non-Mormon miners while the traditional church philosophy continued to influence labor in the Utah coal fields. In the early 1930s, when a delegation of Kenilworth miners requested that UMWA District 22 officials help them organize, they were told that it was useless because they could not get the Mormons to join.[12] The exception to the general church position against labor in the coal fields was that taken by B. H. Roberts in 1933. His death, a few days after his support of the Carbon County miners, was unfortunate for those who hoped that the change in attitude towards organized labor by the federal government also would be reflected by the church.

In summary, the church opposition toward organized labor in the Utah coal fields was influenced by four factors: (1) the general anti-union sentiment that culminated in the unwavering stand for the open shop; (2) the anti-union views of church leaders that had developed from their occupations as businessmen; (3) the threat that the union would create a conflict of loyalties for church members and work to lessen the influence of the church on its members—unionism was fast becoming the "religion of the working class," and church leaders were sensitive to the threat; and (4) the fear that unionzation of the mines would simply raise prices to the Utah consumers, the majority of whom were Mormons.

The attitude of state and local officials towards organized labor in the Utah coal fields played an important part in the failure to establish a permanent and viable union. Although the Utah legislature had looked with considerable favor on the miners through the passage of the eight-hour-day law, through the establishment of a Board of Labor, Conciliation, and Arbitration, through the appointment of a state coal mine inspector in 1896, and through other later measures, the state government, in most cases, served the interests of the coal companies during the several strikes considered in this study.

Even though the policies of Governor Heber M. Wells supporting the coal companies were not as extreme as Colorado's Governor James H. Peabody during the 1903–4 strike, they nevertheless played an important role in influencing public sentiment against the strikers. Wells was branded by the *United Mine Workers Journal* as an "anarchistic Governor" because of his declaration to Charles DeMolli that the union could not win in Utah.[13] The governor was quick to call out the National Guard; and despite the sentiments of many guardsmen in favor of the strikers, the Guard was used to intimidate the strikers, to allow strikebreakers to cross picket lines, and to assist in the eviction of strikers from company property.

During 1922, Governor Charles Mabey showed greater neutrality in handling the strike. He visited the coal fields, met with strikers, and refused to be pushed into calling out the National Guard. When the Guard was finally called out, it attempted to follow a neutral policy for the first few days. Although strikers were forced to give up their weapons, the guardsmen also sought to prevent the entry of strikebreakers into the area. However, this policy changed quickly. Strikebreakers were permitted to go into the mining camps while strikers were prohibited from meeting in them. In the end, the National Guard came to serve the interests of the coal companies as effectively as it had in 1903.

With the exception of Heber M. Wells and Charles Mabey, Utah's governors did not become involved in unionization of the coal fields. Governor Henry Blood left the 1933 strike situation almost exclusively in the hands of local officials and provided only supplemental aid in the form of loaned National Guard weapons and help by a few guardsmen and highway patrol officers. At that time, however, it was not a question of whether organized labor would be recognized in Utah, but rather which of the contending labor unions would speak for the Utah miners.

In reality, local county officers proved more active in their attempts to prevent the organization of the Carbon County area than did state officials. It was they who frantically called on the governor to send in the National Guard during the strikes of 1903 and 1922. The county sheriff was an effective arm of the coal companies as he deputized company guards, arrested union organizers, and added his voice to the call for the National Guard. During the 1912 election in Carbon County, the Socialist and Progressive parties sought to win the votes of the coal miner and railroad worker by denouncing the manner in which earlier county officials had been allied with the coal companies in preventing organized labor from gaining a foothold in the county. Although the Progressive party was partially successful in its reform attempts in Carbon County prior to World War I, the attitude of local authorities toward organized labor was generally at one with the coal interests. In Carbon County, the America labor philosophy of avoiding direct involvement in political affairs proved a detriment. Had labor interests been represented on the county commission, the road to unionization would have been less difficult.

In summary, state and local officials usually came from business backgrounds in which there was little sympathy for organized labor. This and the economic importance of the coal mines to the state and especially Carbon County left little hope that state and local officials would support the goal of unionization in opposition to the interests of the coal companies.

On the two occasions when the federal government became involved in labor affairs in the Utah coal fields, it did so much to the advantage of the labor interests. In 1918 and 1919, the activity of the Federal Mediation and Conciliation Office came closer to insuring that labor would have an unhindered opportunity to organize the Utah miners than any other factor before 1933. Miners, who were threatened with dismissal because of their union affiliation, were protected to some extent by the threat of an investigation and of possible sanctions by the United States Fuel Administration. However, the pressure from the Federal Mediation and Conciliation Office did not last long; and after November 1, 1919, when the nation's coal miners went out on strike despite Woodrow Wilson's action to prevent it, the miners in Utah were left to the mercy of the coal companies. The United States Fuel Company successfully dismissed Frank Bonacci and other leaders and members of the Hiawatha local union of the UMWA. In this case, the Federal Mediation and Conciliation Office did not intervene; earlier in the year the office had acted on behalf of union members dismissed by the Spring Canyon Coal Company for their union membership. In this instance, the national policy of the UMWA worked to the detriment of the nonunion Utah coal fields. Had the union been able to continue its organizing efforts in concert with the policies of the Federal Mediation and Conciliation Office, Utah coal operators might have given up the struggle to prevent unionization of the mines as they did, with relatively little fight, in 1933.

The second time that the federal government directly influenced the course of labor in the Utah coal fields was with the passage of Section 7a of the National Industrial Recovery Act in 1933. That the federal government had openly recognized the right of labor to organize and, by all appearances, openly encouraged unionization placed coal company officials in a strange and uncomfortable position. Unlike the disputes of the past where state and local officials worked hand in glove with the coal companies to stamp out unionization, 1933 saw a new order emerge out of the confusion and frustration of the depression. No other aspect was more important in the recognition of the UMWA in Utah than the change of attitude signalled and momentum generated by the passage of the National Industrial Recovery Act. The influence of organized labor was at its lowest point prior to the passage of the act, and economic conditions offered no opportunity, as they had during World War I, to change the downward slide of organized labor. There is no evidence to indicate that the Utah coal miners would have had another chance to organize until World War II. It is conceivable that the miners may have emerged from the prosperity of the 1940s into the depressed 1950s, as they had from the prosperity of World War I into the 1920s—without union recognition.

In reviewing the overall impact of the federal government on organized labor in Utah, coal miners, who were anxious for union recognition, must be grateful to a government that proved sufficiently flexible to create a new environment without the horror of a violent revolution and civil war.

Utah's citizens were generally opposed to efforts by organized labor to unionize the Utah coal fields. Economically, it was felt that although a union might prove advantageous for the coal miner, it would almost certainly mean higher prices for the consumer. The history of labor strife in the Utah coal fields indicates that attempts to organize were generally followed by an increase in the cost of coal, despite the fact that in most disputes the wages paid the coal miners either remained the same or were decreased. Although the Utah Fuel Company sought to avoid the 1903–4 strike by granting a 10 percent wage increase before the strike was called, the company raised the price of coal by 30 percent to the consumer. In addition, a prolonged strike would greatly reduce the availability of coal, thus driving up the cost above the already inflated level.

Anti-union interests were in large measure successful in branding the attempt to organize the Utah coal miners as being socialistic and communistic inspired. In 1903, it was argued that UMWA organizers like Charles DeMolli and William Price were radicals threatening to destroy the American system and to institute a brand of "foreign socialism" in its place. The same charges were leveled, perhaps with more justification, against the National Miners Union organizers and members thirty years later. Although there was some sentiment for radical change during both strikes, procompany forces were able to play on the general public's fear by proclaiming the threat of organized labor to the American system as greater than it was.

Another facet of the anti-American characteristic of labor, anti-unionists argued, was the strong support by the foreign-born miners.

Before the coal fields were effectively organized by the UMWA, it was announced in every strike that the American miners were showing little interest in the strike while the immigrant miners were responsible for the walkout. The implication was that the American miners were satisfied and had no reason to strike while the "selfish, ungrateful" foreign miner, who was interested only in earning enough money to insure a comfortable living when he returned to his native land, was out on strike, demanding that which loyal citizens did not deem necessary. This, as suggested earlier, prevented many American miners from joining the strikes and in the same manner worked to dampen public sympathy.

A final factor which influenced public opinion against organized labor was the violence which seemed to characterize labor disputes. In Utah, the incidents of violence associated with strikes were relatively few. However, the daily and weekly newspapers, which found their way into the homes of most Utah residents, portrayed a national scene of continuous strife and violence between labor and management that was generally attributed to the uncontrollable and riotous strikers.

To be sure, there were those in the public sector who sympathized with the coal miners and recognized that they were subject to even greater abuse by the coal companies than were the coal consumers. However, with the exception of the Utah Federation of Labor and the Salt Lake Building Trades Council in 1901 and 1903 and of the public protest meetings in support of the National Miners Union in 1933, there was no organized public support for the coal miners during their period of non-unionization. Two newspapers, the *Utah Labor News* and the *Wyoming Labor Journal*, gave some recognition to the plight of the Utah coal miner; however, their readership was relatively small and directed toward those who were already committed to organized labor. On occasion, the Utah daily newspapers showed some sympathy for the coal miner, but their overall posture was decidedly anti-union. On a local basis, the *Eastern Utah Advocate*, and the *Sun*, and *News-Advocate*, (which were later joined as the *Sun-Advocate*) continuously sided with the coal companies during labor disputes and frequently ran articles praising the companies for their benevolent dealings with their employees. The *Carbon County News*, published by W. C. Benfer from 1911 to 1915, supported the goals of unionism and, as mouthpiece for the local Socialist and Progressive parties, challenged the coal companies' political control of the county. However, the *Carbon County News* was published during a period when there was little union activity in the area and was ineffectual as an instrument to promote union organization. The *Helper Journal* was in tune with that community's sentiment for unionization. However, it was only indirectly supportive of the goal of organized labor and did not assume the task of converting nonbelievers to unionism.

The Utah newspapers played a crucial role in shaping public sentiment, and their lack of support for the unionization efforts of the Utah coal miners was of primary importance in the general antagonism of the public during Utah's labor disputes.

Why did it take so long to unionize the Utah coal fields? Were the demands of labor in Utah so radical and unacceptable that they were impossible to achieve? From 1903 until 1933, with the exception of the World War I interlude, the bargaining position of the coal miner steadily worsened. In 1903, the miners struck for a wage increase and union recognition despite a 10 percent pay increase. As the strike continued, union leaders were forced to retreat until the point of contention was whether the Utah Fuel Company would be willing to take back the strikers on an equal basis with the nonunion miners. The company refused, and in the end, the UMWA strike was defeated completely. Although the Western Federation of Miners attempted to pick up the pieces, it failed and unionization efforts ceased until 1918.

The February 1911 strike at Kenilworth was called in protest of company cheating at the weighing scales. It lasted only a few days and did not involve the issue of union recognition. In November 1919, at the height of organizing activity by John McLennan of District 15, the Utah miners did not join the nationwide coal strike and were praised for their conservative and patriotic action. Again in 1922 confronted with another nationwide strike, the Utah miners were excluded from the strike call; however, faced with a 30 percent wage reduction, they chose to strike. According to union leaders, union recognition was not a strike goal as long as members would not be discriminated against because of union affiliation. The strike was basically an attempt to maintain the status quo. Smaller strikes were called during the remainder of the 1920s for the same purpose of preventing wage reductions. However, the strikes were usually unsuccessful, and the reduced wages and infrequent work during the depression left the Utah coal miners in the most unfortunate economic situation in the history of the Utah coal industry. To many, the National Miners Union offered the only possible hope as it attacked the coal company abuses, challenged the political and economic system which had brought the depression, and offered a plan for change. Yet the majority of miners did not join the NMU but instead looked to the more conservative UMWA.

Utah coal miners learned to accept their station and did not readily seek change unless faced with what they considered an intolerable alternative. A few miners subscribed to socialist and communist ideas; however, they quickly learned that if they were to keep their jobs they would have to be careful in expressing their sympathies. The Utah miners did not exhibit the radicalism ascribed to other western labor groups. The conservative nature of their demands and actions and the steadfast manner in which the coal companies refused to consider their modest requests explain in part why the advent of unionism was so slow.

Finally, to explain why success for organized labor was such a long time coming for the Utah coal miners, it is essential to assess the coal miners importance and relationship to the UMWA.

Although there were fewer coal miners in Utah compared with the neighboring states of Wyoming and Colorado, Utah coal found its way into these areas during strikes. The Utah miners received approximately the same pay as the organized miners in Colorado and Wyoming. This situation could be used as an argument against the effectiveness of labor unions. It was an embarrassing situation for the UMWA, epecially during attempts to oust the established organizations.

From 1900 to 1933, the UMWA only took the initiative twice in an effort to organize the Utah miners. In 1903, Utah, as part of District 15, was included in the organizing effort aimed at establishing a union in Colorado. Organizers were sent in and some monetary support was provided to assist those who joined the strike. However, once the strike was lost, the union awkwardly sought to rid itself of the loyal, but burdensome, strikers. This action brought much criticism from the Utah miners, and they turned to the WFM for help in continuing the strike. John Mitchell, it was charged by Utah miners, had sold out to the coal interests. The 1903–4 strike was the longest and most intense before the union success in 1933. The reluctance to meet what the Utah strikers felt was an irrevocable promise of support explains in large measure why there was no strong agitation for affiliation with the UMWA until 1918 and why the Utah miners turned to the WFM.

The second attempt by the UMWA to organize the Utah miners was the campaign directed by John McLennan in 1918 and 1919. Aided by the wartime demand for coal and that somewhat protective environment created by the Federal Mediation and Conciliation Office, McLennan and his deputies succeeded in drawing a significant number of miners into the union. However, with the disagreements between the McLennan-Bonacci factions and the inopportune national strike of November 1919, the organizational effort was inhibited until the change of jurisdiction to District 22 in 1921. The decision by John L. Lewis to grant the Utah request to be assigned to the Wyoming district was wise but might have been more fortunate if it had been made when the change was first requested in June 1920.

The 1922 strike was the most successful during the nonunion period. Although union recognition was not secured, the miners succeeded in preventing the wage reduction that had precipitated the strike. In contrast to 1903, the UMWA did not abandon the Utah miners. At great expense, District 22 officers stood by the miners who were arrested during the strike until they were acquitted or pardoned from prison. As a consequence, the anti-UMWA sentiment that followed the 1903–4 strike did not occur after the 1922 strike. However, union membership quickly declined, and a caretaker arrangement was made to collect, in secret, union dues from old members and recruit new men. During this period, there were no benefits for union membership, and members could be discharged from their employment if their affiliation became known to the company. John L. Lewis considered Tony Radalj, who was responsible for Utah, extravagant, undedicated, and incompetent. As a result of Radalj's inability to handle the Utah situation and of the generally depressed state of the coal industry, unionism completely disappeared from Utah by the late 1920s, except in the dreams of such committed union men as Frank Bonacci.

The UMWA effort in Utah during 1933 was primarily in response to the threat of the National Miners Union. John L. Lewis had instructed District 22 officers to be conservative in the Utah campaign. However, when it appeared that the National Miners Union was on the verge of organizing the entire area, the UMWA jumped in, determined not to lose Utah to the NMU. Although some attempt was made to insist the UMWA had sold out the Utah miners in 1922, the charges were largely ignored. More effective and realistic were the statements by NMU supporters who first approached the UMWA for help in organizing Utah but when turned down in their request turned to the NMU, which willingly took up their cause.

In retrospect, Utah miners more often took the initiative in dealing with their employers to establish the United Mine Workers of America than did UMWA spokesmen. The efforts in Utah by the national organization never followed the idealistic rhetoric of unionism, which honored the attempts by such groups as the UMWA to bring the benefits of union affiliation to all workers as a labor of humane concern. Rather, the UMWA carefully evaluated the importance of Utah to its overall strategy; and when local issues did not mesh with the broader plan, the union was reluctant to commit its resources. Yet, for Utah miners, the dream of organization continued. After it became a reality, Utah miners displayed a fervent loyalty to their hard-won union and to the man they felt had made it possible, Franklin D. Roosevelt.

Appendix

"The Scofield Disaster"

1. Have you heard of the Scofield Disaster?
'Tis a heartbreaking story to tell;
I was there and partook of the sorrow and grief,
I remember, remember so well.
Oh, mothers and wives of the miners,
Who perished so suddenly there,
Did you give them a loving embrace that morn,
Did you bid them 'Goodbye' with a prayer?

Deep, deep down in the mine,
Who knows the danger down in the mine?

2. It was May Day, the year 1900,
The sun shone so bright in the sky;
The miners went merrily downward,
Ne'er dreaming disaster was nigh.

When the mountain it shook and it thundered,
And a blast came from Mine Number Four;
And a hundred and ninety-nine miners,
Lay dead in the terrible roar!

Deep, deep down in the mine,
God, it was awful down in the mine.

3. When the dire news spread through the township,
There went up a heart-rending wail,
As the rescuers brought out the wounded and dead,
And carried them over the trail.

The women and children came running,
To clasp their dear loved ones in death,

Dreadful fire-damp spread from Mine Four up to One,
And the brave rescue men gasped for breath.

Deep, deep down in the mine,
Whose keg of powder blew up the mine?

(Stanzas 4 and 5 missing.)

6. There were hundreds of children made orphans,
And many dear widows were left;
But one poor woman could not be consoled,
She was sadly and madly bereft.

That morning with high temper flying,
Cruel words to her husband she'd spoken;
And so by his graveside she wept and she died,
Anguished, remorseful and broken.

Deep, deep down in the mine,
No man was working down in the mine.

7. You may ask how I know this sad story?
Her five children I took to my home,
Held them close in my arms to still their alarms,
Kept them, and called them my own.

Oh, mothers and wives of the miners,
Strew flowers while life is still fair;
Send your men off to danger with loving embrace,
Bid them 'Goodbye' with a prayer.

Deep, deep down in the mine,
Silence and darkness down in the mine.

This ballad was located and published by LaVerne J. Stallings in *Western Folklore* 18 (1959): 173–76. Permission to reprint from *Western Folklore*. As an introduction to the ballad, Stallings recounts the events that led to the discovery.

> Feeling sure that, like mining disasters elsewhere, this one had been commemorated in balladry, I began to search for evidence in periodicals, histories, and other books in local libraries for some reference to a Scofield ballad. Finding none, I turned to miners and long-time Utah residents that I knew. No one had heard of any verses dealing with the disaster.
>
> My first lead came one evening at a social gathering when, upon telling of my search, a Mr. Larry Kirton said immediately, "Scofield—

why my father was killed in that explosion," and he related the harrowing circumstances. He had heard of a song harshly critical of the mine superintendent, T. J. Parmley, and directed me to a relative of the superintendent, Professor Parmley of the University of Utah. Professor Parmley, when I met him, told me he knew nothing of the ballad, but suggested that I talk with Mr. John Beddoes, one of the few survivors of the disaster. Mr. Beddoes, now 93 and living in Salt Lake City, possessed a most remarkable memory for history and mining songs. He spoke freely of the event, which was still so vivid in his mind that his voice shook and tears rolled down his cheeks as he told how the tragedy had struck almost every family in Scofield. He had known Superintendent Parmley well, he said, and remembered his unusual valor at the time, and that he had been cleared of all blame in the matter. But Mr. Beddoes knew nothing of the ballad in question, nor of any other song reminiscent of the Scofield disaster.

Later, following a lead supplied by Wayland D. Hand, and collaborating with Russell Gray of Ephraim, Utah, I went to Ephraim to seek out an old timer by the name of Arch Draper, who reputedly knew some lines of the Scofield ballad. Ephraim had been my parents' birthplace, and I arrived there with high hopes. Mr. Draper received me cordially and told me that he had done some mining in Scofield years before; while there he had learned a ballad about the disaster. He thought it had been critical of the mine superintendent, but he insisted that he had long ago forgotten the actual words and even the refrain.

Returning to Salt Lake City weary and in low spirits, I unpacked and sat down to rest. This was the former home of my parents where I had been staying during the summer. Along with folklore collecting, it had been my sad task to gather up their last treasured effects, albums, pictures and memoirs, and to sell the house. Just a few minutes after my arrival in late afternoon, a gas company man came to the door to inspect the furnace for prospective buyers. Had I not been there at the moment, what follows might not have taken place. We went downstairs to the basement while thoroughly inspecting the furnace he tore out a partition behind the ducts. There on a shelf was revealed an old metal suitcase, evidently sealed in when the house had been remodeled many years before. Opening this old valise, I found bundles of soot-blackened letters, and underneath them, on the bottom of the case, were five old diaries of Father's, partially discolored and long supposed lost.

I had never seen the diaries before. My previous search forgotten, I sat down immediately and read the engaging account of his life and varied adventures from early boyhood days to young manhood. Father kept excellent accounts. In Volume 5, beginning August 9, 1900, were numerous interesting notations on his mining activities; the last entry summarized the year of 1901, listing his many labors and investments in ten small Utah mines. Beyond the last entry, tucked in between blank pages of the book, lay several small folded papers. I unfolded them. Upon the yellowed pages, and barely discernible, were some penciled verses titled "The Scofield Disaster."

Was this the ballad I had been seeking? It seemingly was not; then one perhaps obscure and unknown? There was no melody, though from the construction it must have been a song. There was no mention of authorship on the pages and two of the numbered stanzas were missing. The diary itself contained no reference to the lines or the disaster. The origin of the song remains a mystery; but perhaps my search will be a guide for other collectors who may be fortunate enough to find someone able to sing the Scofield ballad as recorded in my father's old journal.

The following four untitled poems were written by residents of Springville within a few weeks after the Scofield disaster and published in the *Independent Supplement*, May 24, 1900.

It was a bright spring morning.
The first glad day of May,
When came the sad disaster
That took many lives away.

From the snow-capped peaks of the Wasatsch,
To the valleys that bloom below,
Comes a wail of sadness,
That filled every heart with woe.

Wives weeping for their husbands,
And their sons that are dead.
Oh, such sad bewailing
Many thousand tears were shed.

All hearts were touched with sadness.
But none can never know
Like wives that's lost their husbands,
And their sons that had to go.

God bless the wives and mothers
Where struck the fatal blow!
May he lift the veil of sorrow,
And let peace forever flow.

By Esther Blanchard

Oh! what a solemn sight to see
Father and sons together lie,
All though that great catastrophe
Brother and fathers die.

Those loving friends are called to morn
Their hearts are sad and sore,
But in the resurrection morn
They'll meet to part no more.

Oh, mothers dear do not despair,
Know this your darling boy
Will meet you in another sphere,
And then you will have joy.

Dear children, now your father's dead,
Be to your mothers kind;
Then God will give you daily bread,
A friend in Him you'll find.

Now cease to shed those loving tears.
And now your grief restrain;
The time is drawing very near
When you will meet again.

Oh! Holy Father, hear our prayer,
For 'tis in Thee we trust;
Those mangled bodies lying there
May rise up with the just.

And meet those dear loved ones again,
As they did in days of yore.
Then they will sing the joyful strain,
And live for evermore.

This cruel blow should warn us all
To lead a righteous life
We know not who'll be next to fall—
A husband or a wife.

Parents and children, bear in mind,
We are oft times called to grieve;
Then let us all be good and kind,
And love them while they live.

The heavens wept and tears were shed,
And sorrow came to all

When friends were mingled with the dead
It did strong hearts appall.

In earth's cold tomb our dear ones lay—
They will not long remain;
For in the resurrection day
We'll meet them all again.

Oh God, Thy wisdom we adore,
And may we worthy be,
To live with them that's gone before
Through all eternity.

God bless those great and noble men,
All honor to their name,
Who went into that dangerous glen.
Their deeds deserve great fame.

Who risked their lives in search for friends
To bring them from the mine.
May God to them His power extend,
And give them strength of mind.

Elizabeth Gainge

What means the gloom that rests on every heart,
Which steals the spark of joy from every eye?
What means those clustering groups of earnest men,
Who talk with bated breath and anxious wait
To learn some news they seem afraid to hear?
The air seems thick and laden with woe
That hangs above us like a funeral pall
As if the heavens looked down in sympathy
Upon some general grief too deep for words.
What mean those piteous shrieks, those floods of tears,
Those agonizing moans of hopeless pain,
Whose eloquence affects the hardest heart
And spreads its soft contagion everywhere?
It is the cries of wives for husbands dead,
Of children for the fathers whom they loved,
Of mothers for their missing darling sons,
Of doting sister o'er their brothers gone—
All fallen in that holocaust of death
That happened sudden as a lightning flash
On May the first of this eventful year

So full, so far, of sad calamities.
That morn all went to their accustomed toil
Unconscious of all their approaching fate,
When in an instant came a dreadful sound
And hundreds fell, a suffocated crowd,
All martyrs to the "general public's" needs,
Who surely to the weeping host bereaved
Will graciously extend a helping hand.

Wm. Clegg

Listen to the voice of wailing
Sounding upon the air.
Cries for the loved ones fallen,
In tones of deep despair.
Like a lightning bolt from heaven
That sad disaster fell,
No warning being given,
Few left the tale to tell.

Where once was joy and gladness
There stands a vacant chair,
All mirth is turned to sadness—
The dear one is not there.
Life's weary work is ended,
Its trials all are o'er;
The scenes of earth have faded,
They've gained the other shore.

Oh God, our Heavenly Father,
Who dwells above the sky,
Comfort the lonely widows,
And heed the orphan's cry.
In all their many trials,
Be Thou their hope and stay,
Their shield and their protection,
And guard them night and day.

Mrs. L. H. Mower

Notes

Chapter I

1. The best general study of the Utah labor movement is Sheelwant Bapurao Pawar, "An Environmental Study of the Development of the Utah Labor Movement: 1860–1935" (Ph.D. diss., University of Utah, 1968). Hereafter referred to as Pawar, "The Utah Labor Movement." Pawar's study of the 1890–1920 period was summarized in "The Structure and Nature of Labor Unions in Utah, An Historical Perspective, 1890–1920," *Utah Historical Quarterly* 35 (Summer 1967):236–55. Another leading scholar on Utah labor history is J. Kenneth Davies. His book, *Deseret's Sons of Toil: A History of the Worker Movements of Territorial Utah, 1852–1896* (Salt Lake City, 1977) deals with organized labor in Utah before statehood in 1896. His dissertation, "A Study of the Labor Philosophy Developed Within the Church of Jesus Christ of Latter-day Saints," University of Southern California, 1960, sketches the theological and organizational roots of the Mormon labor philosophy and outlines the history of labor in Utah before providing a detailed discussion of statements by church authorities and the church newspaper, the *Deseret News*, on labor unions and questions relating to labor. His study of the church policy toward the closed shop was published as "Mormonism and the Closed Shop," in *Labor History* 3 (Spring 1962):169–87. An early, but excellent, general study of labor in Utah is Dee Scorup, "A History of Organized Labor in Utah" (M.A. thesis, University of Utah, 1935). Scorup's study is divided into two parts: a history of the entire labor movement in Utah from 1868 to 1934, and a study of the individual unions in the state. A final study dealing with the period between 1910 and 1920 is Karl Alwin Elling, "The History of Organized Labor in Utah 1910–1920" (M.A. thesis, University of Utah, 1962).

2. Pawar, "The Utah Labor Movement," p. 57.

3. Leonard J. Arrington, *Great Basin Kingdom: Economic History of the Latter-day Saints, 1830–1900* (Cambridge, Mass., 1958), p. 333.

4. Pawar, "The Utah Labor Movement," p. 87.

5. Ibid., p. 102, and *Latter-day Saints Millennial Star* 45 (1883):174.

6. Ibid., p. 95.

7. *Deseret News*, October 28, 1885.

8. Ibid., September 8, 1886.

9. Foster Rhea Dulles, *Labor in America: A History* (New York, 1966), pp. 142–47.

10. Pawar, "The Utah Labor Movement," p. 104.

11. Ibid., pp. 106–7.

12. Ibid., pp. 112–14.

13. *Deseret News*, September 7, 1889.

14. Paul A. Frisch, "Labor Conflict at Eureka, 1886-1897," *Utah Historical Quarterly* 49 (Spring 1981): 145–56, and Philip F. Notarianni, "A Tale of Two Towns, Eureka and Helper Utah" (Ph.D. diss., University of Utah, 1978).

15. Pawar, "The Utah Labor Movement," p. 165.

16. Utah, Constitution (1895), Article XVI, Section L.

17. *Holden* v. *Hardy*, 14 Utah 71, 46 p. 756 (1896), affirmed, 169 U.S. 366 (1898).

18. Pawar, "The Utah Labor Movement," p. 190.

19. Ibid., p. 191.

20. Ibid., pp. 211-12.

21. Ibid., p. 207.

22. *Salt Lake Tribune*, September 13, 1911.

23. Pawar, "The Utah Labor Movement," p. 296.

24. Ibid., pp. 296–300, and Helen Zeese Papanikolas, *Toil and Rage in a New Land*, rev. ed. (Salt Lake City, 1974), pp. 121–33.

25. For the definitive work on Joe Hill and his contribution as a songwriter, see Gibbs M. Smith, *Joe Hill* (Salt Lake City, 1969).

26. Ibid., p. 172.

27. Pawar, "The Utah Labor Movement," p. 329.

28. Ibid., pp. 330–31.

29. Ibid., pp. 248–50.

30. *Deseret News*, March 13, 1920.

31. Quoted in Davies, "LDS Labor Philosophy," p. 113.

32. Pawar, "The Utah Labor Movement," pp. 357–58, 367–70.

33. Ibid., p. 384.

34. Ibid., pp. 392–95.

35. Ibid., p. 389.

36. Ibid., p. 401.

37. Ibid., p. 405.

38. Ibid., p. 432.

39. Letter dated July 8, 1849, published in the *New York Tribune*, October 9, 1849, and republished in the *Latter-day Saints Millennial Star* 11 (1849):339.

40. Arrington, "Planning an Iron Industry for Utah 1851–1859," *Huntington Library Quarterly* 21 (May 1958):242.

41. "Manufacture of Iron in Utah," *Latter-day Saints Millennial Star* 17 (1855):5.

42. Gustive O. Larson, "Bulwark of the Kingdom, Utah's Iron and Steel Industry," *Utah Historical Quarterly* 31 (Summer 1963):255–57.

43. *Latter-day Saints Millennial Star* 17 (1855):657.

44. Daughters of Utah Pioneers, *These . . . Our Fathers, A Centennial History of Sanpete County 1849 to 1947* (Springville, Ut. 1947), p. 213, and Alvin Reese Midgely, "A Social Economic Survey of Wales, Sanpete County, Utah." Survey made under the direction of E. C. Bernson, visiting Professor of Rural Social Economics at the Utah Agricultural College, submitted December 15, 1925, included in Wales Ward, Moroni Stake, "Manuscript History," Church of Jesus Christ of Latter-day Saints, Historical Department, Salt Lake City, Utah.

45. Ibid., and Kate B. Carter, "Mining in the West," *Heart Throbs of the West*, vol. 2 (Salt Lake City, 1940), p. 3.

46. Journal History, January 14, 1854, Church of Jesus Christ of Latter-day Saints, Historical Department, Salt Lake City, Utah.

47. Joseph A. Taft, "Notes on the Weber River Coal Field, Utah," State of Utah, *Report of the State Coal Mine Inspector*, 1906, p. 99, State Archives, Salt Lake City, Utah.

48. Marie Ross Peterson, ed., *Echoes of Yesterday* (Ogden, Ut., 1947), pp. 87–88.

49. Hubert Howe Bancroft, *History of Utah* (San Francisco, 1891), p. 737.

50. Peterson, ed., *Echoes of Yesterday*, p. 89.

51. U.S. War Department, *Explorations and Surveys for a Railroad Route to the Pacific Ocean*, vol. 2 (Washington, D. C., 1854), p. 65. The site of the camp was located by C. T. Lupon and described in his "Geology and Coal Resources of Castle Valley in Carbon, Emery and Sevier Counties, Utah," U.S. Geological Survey Bulletin 628 (Washington, D.C., 1916), p. 9.

52. *Salt Lake Tribune*, January 1, 1875.

53. A. C. Watts, "Opening First Commercial Coal Mine Described," in Thursey Jessen Reynolds, comp., *Centennial Echoes From Carbon County* (Price, Ut., 1948), p. 33.

54. Ibid., p. 34. C. H. Madsen presents a somewhat different explanation for the naming of Winter Quarters. He notes, according to the most reliable information available, Winter Quarters was the first coal mine to be opened. "The story has it that a Welsh coal miner led a group of twelve men and one woman across the mountains from Fairview in Sanpete County about 1875. They built a road, opened a small mine and began hauling coal to the settlements in Sanpete County. The first winter was very severe and the small band nearly froze to death. Their camp was pitched in Little Gulch near the place where the Wasatch Store now stands. The town was appropriately named 'Winter Quarters.'" See C. H. Madsen, "Winter Quarters," in Reynolds, comp., *Centennial Echoes from Carbon County*, p. 147. Watts's account seems much more valid since there is no other record of fourteen persons spending the winter in Pleasant Valley, and to open a mine and develop a wagon road between Fairview and Pleasant Valley would have been an enormous accomplishment for the group. In addition the Madsen account does not identify anyone by name—indicating the secondary nature of his source, while the Watts account does provide the names of John Nelson and Abram Taylor.

55. Watts, "Opening First Commercial Coal Mine Described," p. 33.

56. Ibid., p. 35. It is possible that the name Winter Quarters came from the Peter Moran group rather than from John Nelson and Abram Taylor.

57. Winn, "Development of Utah's Coal Mining Industry, 1850–1900," p. 54.

58. Watts, "Opening First Commercial Coal Mine Described," p. 37.

59. Ibid., p. 37. When Watts documented the contribution of the Chinese, he also revealed the pervading nativistic sentiment. "At the reopening of the Utah Mine the company sent in Chinese labor. On their behalf I will say that there is still standing a portion of the mine entry that was driven by them and it is as beautiful a piece of work as one could wish to see in a coal mine. Evidently no powder was used for blasting. The entry was driven exclusively with pick work. The sides are perfectly straight to a certain height and the roof is semi-arched. Due to the method of working, the entry will stand indefinite.

A short time after the Chinese were imported into Pleasant Valley, white labor started to come in and naturally resented the presence of the yellow men. When white labor was strong enough, they brought the situation to a climax and took the law in their own hands. One day they herded the Chinese into a boxcar, fastened the doors, and started the car downgrade. Fortunately, the car kept the track until it reached a place near Hales where there is an adverse grade. It stopped there, and evidently the 'Chinks' traveled the rest of the way on foot. At least they had not been seen in Pleasant Valley from that day to this."

60. Union Pacific Railroad Company, *History of the Union Pacific Coal Mines, 1868 to 1940* (Omaha, 1940), p. 123.

61. Watts, "Opening First Commercial Coal Mine Described," p. 38, and Arrington, *Great Basin Kingdom*, p. 282.

62. Watts, "Opening First Commercial Coal Mine Described," p. 35.

63. State of Utah, *Report of the State Coal Mine Inspector*, 1896, p. 37.

64. Ibid., 1897, pp. 22–23.

65. Ibid., 1901, p. 24.

66. Arthur E. Gibson, "Sunnyside," in Reynolds, comp., *Centennial Echoes From Carbon County*, pp. 203–4.

67. *Eastern Utah Advocate* (Price, Ut.), March 11, 1904.

68. *Deseret News*, December 7, 1906.

69. *Eastern Utah Advocate* (Price, Ut.), March 16, 1905.

70. Ibid., November 14, 1907.

71. Ibid., April 1, 1909.

72. Ibid., July 29, 1909.

73. Journal History, December 6, 1907.

74. See *Salt Lake Herald*, November 21, 1910; November 24, 1910; November 25, 1910; November 29, 1910; December 4, 1910; and December 15, 1910.

75. *Salt Lake Telegram*, November 25, 1910.

76. G. S. Anderson, "History of the Utah Railway Company," *Ax-I-Dent-Ax*, January, 1930.

77. Ibid.

78. Arthur E. Gibson, "The Coal Fields of Eastern Utah, Carbon, and Emery Counties Spring Canyon District Carbon County," in Reynolds, comp., *Centennial Echoes From Carbon County*, p. 224.

79. Production in Storrs was 236,749 tons in 1916 while in Standardville, production was 236,495. Production at the older mines was Winter Quarters, 237,156 tons; Clear Creek, 255,900 tons; Kenilworth, 278,487 tons; Hiawatha, 216,290 tons; and Black Hawk, 274,211 tons. State of Utah, *Report of the State Coal Mine Inspector*, 1916, p. 152.

Chapter II

1. The Scofield explosion ranks as the fourth greatest mine disaster in the United States. Explosions in which loss of life was greater are the following: Monongah, West Virginia, 362 killed on December 6, 1907; Dawson, New Mexico, 263 killed on October 22, 1913; and Jacobs Creek, Pennsylvania, 239 killed on December 19, 1907. U.S. Bureau of Mines, *Historical Summary of Coal-Mine Explosions in the United States, 1810–1958*, by Hiram Brown Humphrey, Bulletin No. 586 (Washington, D.C., 1960), pp. 6, 22–23, 38.

A version of this chapter was published as "Tragedy At Scofield," *Utah Historical Quarterly* 41 (Spring, 1973): 182–94.

2. For examples of poems and ballads about the disaster, see Appendix I.

3. *Salt Lake Herald*, March 23 and 25, 1900.

4. J. W. Dilley, *History of the Scofield Mine Disaster* (Provo, Ut., 1900), pp. 47–48.

5. Ibid., p. 47.

6. *Salt Lake Tribune*, May 4, 1900.

7. *Deseret News*, May 4, 1900, and Dilley, *History of the Scofield Mine Disaster*, p. 56.

8. State of Utah, *Report of the State Coal Mine Inspector*, 1900, p. 62.

9. *Deseret News*, May 8, 1900.

10. Governor Heber M. Wells to Gomer Thomas, May 7, 1900, State of Utah, Industrial Commission, State Coal Mine Inspector's Papers, 1900, State Archives, Salt Lake City, Utah. Mr. Joe Podbevsek, a long-time resident of the Winter Quarters area and acquaintance of many who survived the disaster and participated in the rescue efforts, in a private interview at Scofield, Utah, on December 27, 1971, related that the old-timers claimed that there were as many as one hundred bodies still in the mine. He said that many of the sections of the mine were never worked again. Mr. Podbevsek was convinced from the stories he had heard from the miners who participated in the rescue parties that there are still bodies buried in the mine. Judge Henry Ruggeri, attorney for many of Carbon County's immigrants, in a private interview at Price, Utah, on December 28, 1971, noted that he had heard similar stories from miners who were also involved in the rescue parties. He concluded that it was possible there are still remains of disaster victims in the mine. Others, Tom Biggs, for example, who has lived in Scofield since 1910, claimed that all the bodies were recovered from the mine. Tom Biggs interview, Scofield, Utah, December 27, 1971.

11. Dilley, *History of the Scofield Mine Disaster*, p. 49.

12. *Salt Lake Tribune*, May 4, 1900.

13. Dilley, *History of the Scofield Mine Disaster*, p. 102.

14. Ibid., p. 52.

15. *Salt Lake Tribune*, May 10, 1900.

16. *Salt Lake Herald*, May 7, 1900.

17. For a detailed account of the burial services in these towns, see Dilley, *History of the Scofield Mine Disaster*, pp. 69–90.

18. *Salt Lake Herald*, May 6, 1900.

19. Ibid.

20. *Salt Lake Tribune*, May 4, 1900.

21. Quoted in Dilley, *History of the Scofield Mine Disaster*, p. 206.

22. *Tri-Weekly Journal* (Logan, Ut.), May 5, 1900, and *Salt Lake Herald*, May 6, 1900.

23. State of Utah, *Message of the Governor of Utah to the Fourth Session of the State Legislature of Utah, January 15, 1901*, p. 23.

24. Dilley, *History of the Scofield Mine Disaster*, pp. 218, 220, 222.

25. *Eastern Utah Advocate* (Price, Ut.), May 17, 1900, and *Deseret News*, May 11, 1900.

26. State of Utah, *Message of the Governor of Utah to the Fourth Session of the State Legislature of Utah, January 15, 1901*, p. 23. This total includes the $100,000 donated by the Pleasant Valley Coal Company.

27. Information concerning the age and home districts of the Finns killed in the disaster was provided by Craig Fuller, a student of Finnish history in Utah. Mr. Fuller's source was the May-August issue of the Finnish-language newspaper *Siirtolainen*, published in New York City. Two excellent articles on Finnish immigration are Timo Orta, "Finnish Emigration Prior to 1893: Economic, Demographic and Social Backgrounds," and A. William Hoglund, "No Land for Finns: Critics and Reformers View the Rural Exodus from Finland to America between the 1880s and World War I," in Michael G. Karri, Matti E. Kaups, Douglas J. Ollila, Jr., eds., *The Finnish Experience in the Western Great Lakes Region: New Perspectives* (Turku, Finland, 1975), pp. 21-54.

28. *Salt Lake Tribune*, May 4, 1900, and May 6, 1900.

29. *New York Times*, May 3, 1900.

30. *Deseret News*, May 2, 1900.

31. D. J. Parker, "Explosions in Utah Coal Mines," U.S. Bureau of Mines, Information Circular 6732 (Washington, D.C.), pp. 5, 7.

32. See State of Utah, *Report of the State Coal Mine Inspector*, 1900, pp. 64–97, and *Message of the Governor to the Fourth Session of the State Legislature of Utah, January 15, 1901*.

Chapter III

1. *Intermountain Catholic* (Salt Lake City), May 5, 1900.

2. *Deseret News*, January 24, 1901.

3. The members of this committee were Pat Rooney, George Lewis, Bert Sinden, Andrew Smith, John Bell, John Samuels, Robert Walker, and one Italian and one Finn selected to represent their nationalities. *Salt Lake Herald*, January 14, 1901.

4. Ibid., February 1, 1901. In a communication of January 21, 1901, the committee had stated:

"The 12th day of January of this new year found both employers and employees in comparative unity and harmony, which had also been the case for a long period prior to that, but on account of the general prosperity of the past year, and especially that of the coal trade in which the prices of coal so much advanced, and that so rapidly, the employees of the said company began to think that a small share of the profits would be just and in order, to which end and purpose a little wind of agitation blew up among them.

A mass meeting, however, was at once called, and after due consideration, we came to the unanimous conclusion that we ask an advance of 15 cents per ton for the coal mined, and a proportionate advance on all day labor." *Deseret News*, January 24, 1901.

5. *Salt Lake Tribune*, January 14, 1901.

6. Ibid.

7. *Salt Lake Herald*, January 14, 1901.

8. Ibid.

9. Members of the committee included: Chairman J. C. Bell, Robert Walker, John Samuels, Joseph Bronick, John Landine, George Lewis, P. J. Rooney, Bert Sinden, all of Scofield; as well as Thomas F. Hoye, William Eynon, Fred Wolf, John Hendrickson, Guy Castalino, Frank Plastino, and Leo Hampton, all of Clear Creek.

10. W. G. Sharp in a letter, apparently intended for the public, wrote to H. G. Williams: "We consider that the company has treated the Winter Quarters men fairly. They are making good wages, and further, we consider that the company was not treated fairly when the men made their preemptory demands, refusing even to give us three days' time to discuss and consider the questions raised. By their own actions and under present conditions they have deliberately cut off discussion and prevent the company from considering or taking any action." Quoted in *Deseret News*, January 25, 1901.

11. Ibid., February 5, 1901.

12. For the text of the miners' letter, see *Salt Lake Tribune*, January 30, 1901.

13. Ibid., February 3, 1901. The company reported the miners were earning between $3 and $5 a day. The strikers placed their daily earnings at between $1.50 and $2 a day.

14. Ibid., February 17, 1901.

15. Statement to the public by the miners' committee, J. C. Bell, Lewis George, and Pat Rooney, quoted in *Salt Lake Tribune*, January 30, 1901.

16. This problem on tonnage was discussed by the miners' committee in an article appearing in the *Deseret News*, January 24, 1901. "It would be well for the public to know our demands are naught but reasonable, since we have to load not less than 3,200 pounds of coal for 60 cents; 2,240 pounds is the ton in vogue here for 30 percent is taken off or deducted for slack so that if we load on one car 2,240 pounds or one ton, we only get paid for 1,568 pounds, which is about 42 cents, while the company in selling give, 2,000 pounds for a ton, thus gaining in the transaction 1,200 pounds per ton, together with the market price on the ton besides. . . ."

17. *Deseret News*, January 30, 1901.

18. Ibid.

19. Ibid.

20. *Salt Lake Herald*, February 1, 1901.

21. The Building Trades Council was established on May 1, 1899. The affiliated unions included the Bricklayers; Tin, Sheet Iron and Cornice Workers; Carpenters and Joiners; Stonecutters; Plasterers; and Painters. Pawar, "The Utah Labor Movement," pp. 162, 179.

22. The Utah Federated Trades and Labor Council, predecessor of the Utah Federation of Labor, was organized in 1889. It, as well as the Utah Federation of

Labor, was made up of organized trade unions such as the Cigarmakers, Typographers, Blacksmiths, Horseshoers, Streetcarmen, Barbers, Brewers, Brickmakers, Machinists, Tailors, Locomotive Engineers, Locomotive Firemen, and Railroad Trainmen. Ibid., pp. 107, 109, 125, 155, 156.

23. *Salt Lake Herald*, January 28, 1901, and *Salt Lake Tribune*, January 29, 1901.

24. State of Utah, *Report of the State Coal Mine Inspector*, 1903, p. 77.

25. *Salt Lake Tribune*, February 8, 1901, reported: "The Italians, it is stated, said openly they would be with the strikers to a man if they dared."

26. *Salt Lake Tribune*, February 4, 1901.

27. *Salt Lake Herald*, February 5, 1901.

28. Ibid., February 8, 1901.

29. *Eastern Utah Advocate* (Price, Ut.), February 14, 1901, see also Notarianni, "Italian Fraternal Organizations in Utah, 1890–1933," *Utah Historical Quarterly* 43 (Spring 1975): 172–87.

30. *Eastern Utah Advocate* (Price, Ut.), February 14, 1901.

31. Because of the distance from Scofield to Sunnyside, approximately seventy-five miles, the general fatigue of the strikers from the exposure and long marches, and the failure at Castle Gate, no concerted effort was made to carry the strike to Sunnyside.

32. *Salt Lake Herald*, February 3, 1901. There was also some speculation that John Mitchell, president of the United Mine Workers of America, would come to Utah. See *Salt Lake Herald*, February 15, 1901.

33. *Eastern Utah Advocate* (Price, Ut.), March 28, 1901.

34. *Salt Lake Tribune*, February 4, 1901.

35. *Eastern Utah Advocate* (Price, Ut.), February 7, 1901.

36. *Salt Lake Herald*, February 16, 1901.

37. *Salt Lake Tribune*, February 4, 1901.

38. "Castle Gate miners are largely foreigners and it was this element that forced the more conservative at Clear Creek and Winter Quarters to quit work. . . ." Ibid., January 31, 1901.

39. *Deseret News*, January 24, 1901.

40. Ibid., February 9, 1901.

41. This was reported by striker Bert Sinden and quoted in the *Salt Lake Herald*, February 16, 1901.

42. Ibid., January 30, 1901.

43. *Salt Lake Tribune*, January 14, 1901.

44. *Deseret News*, February 9, 1901.

45. *Salt Lake Tribune*, February 23, 1901.

46. For a comparison of these laws, see Title 36, Chapter 1, *The Revised Statutes of the State of Utah*, 1898, pp. 354–55, and Chapter 68, *Laws of the State of Utah Passed at the Fourth Session of the Legislature of the State of Utah*, pp. 68–70.

47. Chapter 44, *Laws of the State of Utah Passed at the Fourth Session of the Legislature of the State of Utah*, p. 39.

48. *Deseret News*, January 24, 1901.

49. *Salt Lake Tribune*, January 25, 1901.

50. *Eastern Utah Advocate* (Price, Ut.), February 14, 1901.

51. *Salt Lake Tribune*, February 1, 1901. There were several other "lawless" acts, but they were not necessarily a result of the strike. An Italian track walker was held up between Clear Creek and Winter Quarters by five men. When they found he had no money they left him alone. Another man by the name of Schwartz, traveling between Winter Quarters and Scofield, was knocked unconscious and robbed. Ibid., January 24, 1901. At Scofield a Finn was arrested at the instigation of a young woman who claimed he had insulted her. The man was found guilty and in an interesting example of coal-camp justice was required to pay the woman $50. *Eastern Utah Advocate* (Price, Ut.), February 14, 1901.

52. *Salt Lake Herald*, February 16, 1901, and *Salt Lake Tribune*, February 17, 1901.

53. *Salt Lake Tribune*, February 17, 1901.

54. Ibid.

55. *Deseret News*, February 4, 1901.

56. *Salt Lake Herald*, January 30, 1901.

57. Ibid., February 12, 1901.

58. *Salt Lake Tribune*, February 11, 1901.

59. *Salt Lake Herald*, February 11, 1901.

60. Ibid., February 16, 1901.

61. Ibid., February 17, 1901.

62. *State of Utah, Report of the State Coal Mine Inspector*, 1901, p. 19. The February 28, 1901 issue of the *Eastern Utah Advocate* (Price, Ut.) reported that approximately 150 men were not rehired because they refused to sign the "ironclad." Mine Inspector Gomer Thomas summed up the strike situation in these words: "From the 13th of January to the 21st of February, 1901, the mine at Winter Quarters was idle on account of a strike, in which the employees wanted an increase in their wages. At the end of 38 days they all went back, or nearly all. Those who took part in the strike were refused work and had to leave camp, and their places were filled with foreign labor, which was a great drawback to the management, as the new men could not produce the amount of coal, nor quality. This strike was like all others—both sides were losers, the employees having the worst of it." See State of Utah, *Report of the State Coal Mine Inspector*, 1901, p. 17.

63. *Salt Lake Tribune*, February 26, 1901.

Chapter IV

1. A condensed version of this chapter was published as "The 'Foreign Element' and the 1903–1904 Carbon County Coal Miners Strike," *Utah Historical Quarterly* 43 (Spring 1975):125–54.

2. Selig Perlman and Philip Taft, *History of Labor in the United States, 1896–1932*, 4 vols. (New York, 1935), 4:13. However, it should be noted that during this time less than 10 percent of the nation's wage earners were involved in the unions as members. Dulles, *Labor in America*, p. 186.

3. The strikers' demands were reported in a letter from union representatives Thomas Phelps, Sunnyside; Louis Jones, Winter Quarters; Thomas Potter, Clear

Creek; and Thomas Lewis, Castle Gate; to G. W. Kramer, vice president of the Utah Fuel Company, quoted in the *Salt Lake Tribune*, December 20, 1903, p. 18, and in State of Utah, *Report of the State Coal Mine Inspector*, 1903, pp. 63–64.

4. *Salt Lake Tribune*, November 19, 1903.

5. *Salt Lake Herald*, November 26, 1903.

6. In a letter from John Mitchell, president of the United Mine Workers of America, to Edwin F. Holmes, president of the Salt Lake City Commercial Club, November 27, 1903, Mitchell denied that the Utah strike had been called as a sympathetic strike for the Colorado strikers. "Utah miners are not striking in sympathy with miners of Colorado. On the contrary, they demand an increase of 20 percent in wages. Not to exceed three organizers are stationed in Utah. It is unreasonable to suppose that two or three organizers could intimidate Utah miners and prevent their working if they desired. I am anxious for an amicable settlement and am willing to meet representatives of the coal company for the purpose of arranging conditions of settlement." Quoted in the *Salt Lake Tribune*, November 28, 1903.

7. T. Kaisley, secretary of the Scofield Local of the UMWA, to the editor of the *United Mine Workers Journal*, February 28, 1904.

8. Notarianni, "Italian Involvement in the 1903–1904 Coal Miners' Strike in Southern Utah and Utah," Unpublished Seminar Paper, Department of History, University of Minnesota, June 1975, p. 16.

9. According to the *Eastern Utah Advocate* (Price, Ut.), December 3, 1903, the mining population of the Carbon County camps was as follows:

Camp	Italian	English Speaking	Austrian	Finn
Castle Gate	356	108	10	
Sunnyside	246	356	222	
Clear Creek	172	95		128
Winter Quarters	74	181		126

10. See State of Utah, *Report of the State Coal Mine Inspector*, 1903, p. 47, for the text of the letter dated November 21, 1903, from Hyrum Wilcox to Governor Heber M. Wells.

11. The parades were colorful affairs. On February 18, 1904, the *Eastern Utah Advocate* (Price, Ut.) reported that a parade had been held at Helper in which the strikers masqueraded as capitalists, miners, and slaves.

12. *Salt Lake Herald*, December 11, 1903.

13. Reports that strikers intimidated miners were sent by H. G. Williams to Adjutant General Charles S. Burton. See State of Utah, Industrial Commission, State Coal Mine Inspector's Papers, "Report—Militia At Scofield During Miners Strike, 1904."

14. *Salt Lake Telegram*, December 2, 1903.

15. *Deseret News*, November 24, 1903.

16. Lott Powell, one of the guards at Sunnyside, said that although he agreed with the miners' grievances, the unbelievable pay offered men to become company guards could not be turned down. Lott Powell interview, Salt Lake City, Utah, January 3, 1971.

17. One such guard, C. L. "Gunplay" Maxwell, who continually sought to become a bona fide member of the Robber's Roost gang, was the antithesis of the gallant Butch Cassidy. One story from eastern Utah tells how Gunplay Maxwell rode

into a camp of prospectors who were looking for gilsonite in the Hill Creek country east of Green River. To demonstrate how bad he was, Maxwell tried to pick a fight with one of the smaller men. The prospectors invited Maxwell to take off his guns and fight. He boastfully replied, "I never take my guns off." "The hell you don't!" shouted the prospectors as they stripped Maxwell of his guns and turned the man loose on him. Following the fight, the half-dead badman was loaded on his horse and sent on his way. In 1898 Maxwell had been sentenced to prison for eighteen years for robbing the Springville bank, but on November 21, 1903, his sentence was commuted by influential friends, presumably Utah Fuel Company officials, and he went to Carbon County as a company guard. Charles Kelly, *The Outlaw Trail* (New York, 1959), p. 185.

18. *United Mine Workers Journal*, January 21, 1904.

19. For the text of General Cannon's report, see State of Utah, *Report of the State Coal Mine Inspector*, 1903 and 1904, pp. 50–56.

20. For an account of the role of the National Guard in the strike, see Richard Campbell Roberts, "History of the Utah National Guard: 1894–1954" (Ph.D. diss., University of Utah, 1973), pp. 58–72.

21. Matt T. Howard, interview, Nephi, Utah, February 4, 1971. Howard was a member of the Nephi Company called out in 1903.

22. Citizens of Sunnyside, including the strikers, were quite hospitable to the troops. "Whenever a soldier enters the hut of a miner, wine, beer, and whiskey is at once produced and offense is taken if their hospitality is not accepted." *Salt Lake Tribune*, December 29, 1903.

23. *Nephi Record*, December 11, 1903.

24. *United Mine Workers Journal*, February 28, 1904.

25. *Salt Lake Herald*, December 7, 1903, and December 8, 1903.

26. *Salt Lake Tribune*, November 24, 1903.

27. Matt T. Howard interview.

28. Ibid.

29. George Graham Suggs, Jr., "Colorado Conservatives Versus Organized Labor: A Study of the James Hamilton Peabody Administration 1903–1905" (Ph.D. diss., University of Colorado, 1964), p. 490. *Il Lavoratore Italiano* (Trinidad, Co.), March 14, 1904.

30. Governor Heber M. Wells to General John Q. Cannon, December 3, 1903, "Letterbook of Heber M. Wells, Governor," State Archives, Salt Lake City, Utah.

31. The report was published in the *Deseret News*, December 8, 1903.

32. *Salt Lake Tribune*, December 8, 1903.

33. *United Mine Workers Journal*, December 10, 1903.

34. *Salt Lake Herald*, December 8, 1903.

35. W. Lloyd Adams, "Memoirs of W. Lloyd Adams," pp. 6–7. A typescript copy of this portion of his unpublished memoirs was obtained by the author from Adams's niece Sadie Greenhalgh, Nephi, Utah.

36. A.B. Edler, "Champion of the New Civilization," Handbill, Utah State Historical Society, Salt Lake City, Utah.

37. *Salt Lake Tribune*, December 11, 1903.

38. Cannon's report was published in the *Deseret News*, December 14, 1903.

39. *Salt Lake Herald*, December 14, 1903.

40. Ibid., December 25, 1903.

41. *Salt Lake Tribune*, December 23, 1903.

42. *Salt Lake Herald*, November 25, 1903.

43. Ibid., November 20, 1903.

44. *Deseret News*, December 21, 1903.

45. Ibid.

46. Ibid., December 7, 1903.

47. Quoting from a letter written by General John Q. Cannon to Adjutant General Charles S. Burton, December 14, 1903: "The two companies here [Castle Gate] are so widely separated as to their camps that it is found necessary to have them mess separately. . . .

"Another reason for causing the two companies to mess separately I dislike extremely to mention, but feel that perhaps I should do so. It is in brief that quite a number of the members of 'H' company are found to be lousy. We are taking the most heroic measures to eradicate the evil, and upon the receipt of certain supplies tomorrow shall thoroughly fumigate all bedding and tentage; have the men thoroughly bathed and put into new clothing, which at their expense we have been able to get from the store here. The first indication of the most lamentable condition was noted two or three days ago in one man who was immediately isolated from the others; but it has spread with considerable rapidity, though the company itself is made up of men who are on the whole rather superior as to their habits of cleanliness." See "Report—Militia At Scofield During Miners Strike, 1904."

48. *Salt Lake Tribune*, December 18, 1903.

49. Matt T. Howard interview.

50. Ibid.

51. *Deseret News*, December 8, 1903.

52. *Salt Lake Tribune*, December 14, 1903. Commenting on this, the newspaper reported: "The public has no interest whatever in supporting a blacklist against union men; and the time has come when the public must be considered, above any claim or contention of either party to the dispute."

53. *Salt Lake Herald*, December 20, 1903.

54. State of Utah, *Report of the State Coal Mine Inspector*, 1903, p. 68.

55. Samuel A. King, "Statement and Brief Concerning the Campaign of the Coal Operators in Utah against Organized Labor and the Unionizing of the Utah Coal Fields," Photocopy, Utah State Historical Society, Salt Lake City, Utah.

56. Some Carbon County citizens felt that the evacuation of the National Guard was premature. Strikers were among those who requested that the militia remain. "The strikers themselves, in the form of a petition in one instance, and in repeated personal requests to and conversations with me urged that the troops remain. It was peculiarly gratifying to feel that in a situation of such delicacy, our soldiers had gained the confidence of all elements of the community." Cannon to Burton, December 31, 1903, in "Report—Militia At Scofield During Miners Strike, 1904."

57. *Deseret News*, December 4, 1903.

58. *Salt Lake Tribune*, December 29, 1903.

59. *United Mine Workers Journal*, January 21, 1904.

60. Ibid., September 8, 1904.

61. *Deseret News*, January 4, 1904.

62. H. G. Williams to Joseph F. Smith, December 2, 1903, Joseph F. Smith Papers, Church of Jesus Christ of Latter-day Saints, Historical Department, Salt Lake City, Utah.

63. Joseph F. Smith, John R. Winder, and Anthon H. Lund to Reuben G. Miller, December 4, 1903, First Presidency Letterpress Books, Church of Jesus Christ of Latter-day Saints, Historical Department, Salt Lake City, Utah.

64. Joseph F. Smith, John R. Winder, and Anthon H. Lund to H. G. Williams, January 25, 1904, First Presidency Letterpress Books, Church of Jesus Christ of Latter-day Saints, Historical Department, Salt Lake City, Utah.

65. Helen Zeese Papanikolas, *Toil and Rage in a New Land*, pp. 109–10, and *Eastern Utah Advocate* (Price, Ut.), April 21, 1904.

66. *Eastern Utah Advocate* (Price, Ut.), March 3, 1904.

67. *Deseret News*, December 8, 1903.

68. Ibid., April 30, 1904.

69. Ibid.

70. Ibid.

71. Dr. J. J. Dalpiaz in an interview with Philip Notarianni and Kent Powell at Helper, Utah, February 5, 1972, gave the following statement about how his father, Celeste Dalpiaz, and another Italian, Angelo Pilatti, regarded the burning of the shed: "Later when Mother Jones, the old union organizer, came through the country down here they were all going to lock her up. They built a house right over here right toward the east side of Helper over here. On the big flat and they were going to lock her up in this house. My dad and a fellow by the name of Angelo Pilatti walked over the hill way back there. They came over that hill way back and came down. They both had a gallon of kerosene. They both came down and went over that place. The shingles were not dry. They were still kind of green and it was wet and they couldn't. . . . They spread this thing all over and they set the thing on fire and then they went over the hill down in and they went around Helper and went back. But my dad says by the time we got to the hill it was burning pretty good, but he said by the time we got on top it was smoldering out. They ruined it so that it couldn't be used."

72. Mother Jones, *Autobiography of Mother Jones*, ed. Mary Field Barton (Chicago, 1925), pp. 105–6.

73. *Deseret News*, April 25, 1904.

74. Lott Powell interview.

75. *Salt Lake Herald*, April 27, 1904, and *Deseret News*, April 26, 1904.

76. *Salt Lake Herald*, April 26, 1904.

77. However, this was not the end of the episode. A few days after his release, Joseph Barboglio, treasurer of the United Mine Workers of America and later founder of the Helper State Bank and the Carbon Emery Bank, brought a $5,000 suit against Wilcox. The charge was false imprisonment. Apparently the suit was either dismissed or dropped because no further mention was made of the case in newspapers or court records.

78. Newell Christy Remington, "A History of the Gilsonite Industry" (M.A. thesis, University of Utah, 1959), pp. 36–40.

79. *Salt Lake Herald*, April 25, 1904.

80. In a letter to Harry Bousfield, President John Mitchell admitted union defeat: "Close inquiry upon the part of our representatives satisfied us that the mines are being operated with reasonable success and there is no possibility of winning the strike. For this reason we have decided to bring the strike to a close at the earliest possible date. . . . the strike in District 15 has been more expensive to the national organization than any other movement in our history in which a like number of men have been involved. . . . It is our sincere hope that all those who have participated in the struggle may be able to secure speedy employment, that they may retain their interest in the organization and prepare themselves for a more successful movement at the first opportunity. It is not our intention to abandon the work of organization in Colorado and Utah. We propose continuing our efforts to organize the men of these fields." Quoted in U.S. Congress, Senate, Commissioners of Labor, *A Report of Labor Disturbances in the State of Colorado, from 1880 to 1904, Inclusive with Correspondence Relating Thereto*, Doc. No. 122, 58th Cong., 3d sess., 1904–1905 (Washington, D.C., 1905), pp. 356–57.

81. *Eastern Utah Advocate* (Price, Ut.), June 2, 1904.

82. *United Mine Workers Journal*, August 11, 1904.

83. Marchiori's report was published in an Italian newspaper, published in San Francisco, *La Protesta Umana*, April 9, 1904. Quoted in Notarianni, "Italian Involvement in 1903–1904 Coal Miners' Strike in Southern Colorado and Utah," pp. 36–37.

84. "Verbatim Report of the Proceedings relative to Charges Made by Robert Randall," *United Mine Workers Journal*, February 9, 1905.

Chapter V

1. Dale Featherling, *Mother Jones The Miners' Angel: A Portrait* (Evansville and Carbondale, Ill., 1974), pp. 59–60.

2. *Eastern Utah Advocate* (Price, Ut.), March 9, 1905.

3. Ibid., June 1, 1905, and *Official Proceedings of the Thirteenth Annual Convention, Western Federation of Miners, May 22 to June 9, 1905*, Denver, Colorado, 1905, p. 213.

4. *Official Proceedings of the Fourteenth Annual Convention, Western Federation of Miners, May 28 to June 13, 1906*, Denver, Colorado, 1906, pp. 59–62, 64.

5. John E. Brinley, "The Western Federation of Miners" (Ph.D. diss., University of Utah, 1972), pp. 223–24.

6. *Official Proceedings of the Nineteenth Annual Convention, Western Federation of Miners, July 17 to August 4, 1911*, Denver, Colorado, 1911, p. 200.

7. *Eastern Utah Advocate* (Price, Ut.), June 1, 1905.

8. Ibid., June 14, 1906.

9. *Official Proceedings of the Fifteenth Annual Convention, Western Federation of Miners, June 10 to July 3, 1907*, Denver, Colorado, 1907, pp. 92–93.

10. *Eastern Utah Advocate* (Price, Ut.), March 9, 1905.

11. Ibid., November 2, 1905.

12. Ibid., March 7, 1907.

13. Ibid., July 11, 1907.

14. Ibid., October 31, 1907

15. Ibid., September 5, 1907.

16. Ibid., January 2, 1908.

17. Ibid., January 9, 1908, p. 3, and *Official Proceedings of the Sixteenth Annual Convention, Western Federation of Miners, July 13 to July 29, 1908*, Denver, Colorado, 1908, p. 289.

18. Ibid., p. 295.

19. Ibid.

20. Helen Zeese Papanikolas, *Toil and Rage in a New Land*, pp. 104–5.

21. Ibid., p. 110, and Helen Zeese Papanikolas, *The Peoples of Utah* (Salt Lake City, 1976), pp. 410–11.

22. *Sun* (Price, Ut.), April 9, 1915.

23. Helen Zeese Papanikolas, *The Peoples of Utah*, p. 412.

24. The unsigned letter was written February 12, 1911, from Bingham and addressed to "Utah Government, Salt Lake City, Utah." The letter is in the William Spry Correspondence, 1910–1912, Box 10, State Archives, Salt Lake City, Utah.

25. Letter from R.C. Gemmell to Governor William Spry, July 23, 1912, Ibid., Box 18.

26. For an account of the 1912 Bingham Strike, see Helen Zeese Papanikolas, "The Great Bingham Strike of 1912 and Expulsion of the Padrone," *Toil and Rage in a New Land*, pp. 121–33.

27. *Eastern Utah Advocate* (Price, Ut.), June 25, 1908.

28. Ibid., June 18, 1908.

29. Ibid., July 30, 1908.

30. Ibid., October 28, 1908.

31. *Sun* (Price, Ut.), April 16, 1915.

32. Ibid., July 30, 1915.

33. The letter was quoted by J.E. Pettit, State Coal Mine Inspector, in his report about the Kenilworth trouble to Governor William Spry, February 11, 1911. William Spry Correspondence, 1910–1912, Box 9, State Archives.

34. *Salt Lake Herald*, February 7, 1911.

35. Ibid.

36. *Salt Lake Tribune*, February 8, 1911.

37. *Salt Lake Herald*, February 7, 1911.

38. *Deseret News*, February 8, 1911.

39. Pettit to Spry, February 11, 1911, William Spry Correspondence, 1910–1912, Box 9, State Archives.

40. *Deseret News*, February 7, 1911.

41. *Salt Lake Herald*, February 8, 1911.

42. *Salt Lake Tribune*, February 8, 1911.

43. *Deseret News*, February 8, 1911.

44. *Eastern Utah Advocate* (Price, Ut.), February 16, 1911.

45. Ibid., March 16, 1911, and *Salt Lake Herald*, March 15, 1911.

46. *Carbon County News* (Price, Ut.), September 5, 1912.

47. Ibid. The Socialist county ticket included state legislature representative, George W. Rauch; four-year county commissioner, Charles Rugerri; two-year county commissioner, K. A. Thompson; sheriff, W. A. Jameson; county treasurer, I. W. Stark; county clerk, Aubrey Roberts; county recorder, Harmon Stoltz; county attorney, L. C. Hall; county assessor, James Dunton; county surveyor, Peter Anderson; superintendent of schools, Mrs. Archie Allison.

48. Ibid.

49. Ibid., October 24, 1912.

50. Ibid., October 31, 1912.

51. For a discussion of the Progressive attacks on the Utah Fuel Company, *Eastern Utah Advocate* (Price, Ut.), and Mark Braffett, see Lynda Broadbent Gibbs, "The History of the United Mine Workers of America in Carbon County, Utah, Up to 1933" (M.A. thesis, Brigham Young University, 1968), pp. 64–75.

52. *Carbon County News,* January 30, 1913.

53. Ibid., April 10, 1913.

54. Joyce L. Kornbluh, ed., *Rebel Voices, An I.W.W. Anthology* (Ann Arbor, Mich., 1968), p. 1.

55. *Salt Lake Tribune,* June 13, 1913.

56. Ibid., June 12, 1913.

57. Ibid.

58. Ibid., June 13, 1913.

59. *Salt Lake Herald,* June 14, 1913.

60. Ibid., June 13, 1913.

61. *Salt Lake Tribune,* June 12, 1913.

62. *Salt Lake Herald,* June 12, 1913.

63. *Eastern Utah Advocate* (Price, Ut.), June 19, 1913.

64. *Salt Lake Herald,* June 14, 1913.

65. Ibid., August 13, 1913, and *Salt Lake Tribune,* August 13, 1913.

66. *Eastern Utah Advocate* (Price, Ut.), August 21, 1913.

67. For an account of Joe Hill, his trial and execution in Utah, see Smith, *Joe Hill.*

68. For an account of the 1913–14 Colorado Strike, see George S. McGovern and Leonard F. Guttridge, *The Great Coalfield War* (Boston, 1972); Barron B. Beshoar, *Out of the Depths, The Story of John R. Lawson* (Denver, 1958); and Zeese Papanikolas, *Buried Unsung: Louis Tikas and the Ludlow Massacre* (Salt Lake City, 1982).

69. J. E. Pettit to William Spry, September 29, 1913, William Spry Correspondence, 1913, Box 26, State Archives.

70. *Eastern Utah Advocate* (Price, Ut.), November 13, 1913.

71. Ibid., October 2, 1913.

72. William Littlejohn to J. E. Pettit, September 29, 1913, State of Utah, Industrial Commission, State Coal Mine Inspector's Papers, 1913.

73. *Eastern Utah Advocate* (Price, Ut.), May 14, 1914.

74. William Bell to J. E. Pettit, March 1, 1915, State of Utah, Industrial Commission, State Coal Mine Inspector's Papers, 1915.

75. Ibid., April 10, 1915.
76. Ibid., May 26, 1915.
77. Ibid., September 9, 1915.
78. Ibid., October 13, 1915.

Chapter VI

1. *Salt Lake Tribune*, June 19, 1917.
2. W.C. Ferguson, General Superintendent of the Calumet Fuel Company, to A.H. Cowie, April 13, 1917, Utah Fuel Company Records, MSS 154, Box 15, Brigham Young University Library, Provo, Utah.
3. Telegram from A.H. Cowie, April 12, 1917, Utah Fuel Company Records, MSS 154, Box 15, Brigham Young University Library, Provo, Utah.
4. *Salt Lake Herald*, February 14, 1917.
5. *Sun* (Price, Ut.), February 16, 1917.
6. Ibid.
7. Ibid., February 23, 1917.
8. *Deseret News*, February 23, 1917.
9. *Sun* (Price, Ut.), June 22, 1917.
10. Ibid., September 7, 1917.
11. Ibid., September 14, 1917.
12. Correspondence signed by John P. White, president; Frank J. Hayes, vice president; and William Green, secretary-treasurer, United Mine Workers of America, February 8, 1917, and addressed to the Officers and Members of all Local Unions, District 15, United Mine Workers of America, United Mine Workers of America Records, District 15 Correspondence, 1917, UMWA Headquarters, Washington, D.C.
13. Between October 18, 1918, and April 14, 1919, the following men were paid to assist the organizing work: Dominic Bianco, $8; William Knight, $25; Geo. Baca, $6; Joe Prezel, $10; Robert Taurmani, $16; E.H. Clair, $14; Jas. Dorun, $155; Gust. Gracco, $10; M. Bergera, $16; Henery Sweet, $25; J.K. Ramer, $20; S. Lewis, $20; Geo. Pappas, $15; Gust. Drosser, $14; Harry G. McConnell, $10; and J.O. Edwards, $16. "Statement of Funds Received From International Secretary William Green for the State of Utah." Submitted by John McLennan, June 10, 1919, United Mine Workers of America Records, International Executive Board, File Documents and Circulars, June 1, 1919—June 1, 1920.
14. Statement made by John McLennan, International Representative, United Mine Workers of America, to the Industrial Commission of Utah. Records of the Federal Mediation and Conciliation Service, Record Group 380 (33-2496), National Archives, Washington, D.C.
15. Quoted in *Wyoming Labor Journal*, December 6, 1918.
16. J.A. Harvey telegram to William B. Wilson, October 11, 1918, Records of the Federal Mediation and Conciliation Service, Record Group 380 (33-2496), National Archives.
17. A.E. Harvey, secretary of the Utah Federation of Labor, telegram to William B. Wilson, October 5, 1918, Records of the Federal Mediation and Conciliation Service, Record Group 380 (33-2453), National Archives.

18. Preliminary Report submitted by James Doyle, December 8, 1918, Records of the Federal Mediation and Conciliation Service, Record Group 380 (33–2496), National Archives.

19. R.W. Van Derck correspondence, Records of the Federal Mediation and Conciliation Service, Record Group 380 (33–2496), National Archives.

20. William M. Knerr, industrial commissioner of Utah, to W.B. Wilson, secretary of labor, October 10, 1918, Records of the Federal Mediation and Conciliation Service, Record Group 380 (33–2496), National Archives.

21. *Sun* (Price, Ut.), January 17, 1919.

22. *Utah Labor News* (Salt Lake City), January 17, 1919. This copy was found in the Records of the Federal Mediation and Conciliation Service, Record Group 380 (33–2496), National Archives.

23. *Sun* (Price, Ut.), January 17, 1919.

24. Charles T. Connell, "United Mine Workers and Spring Canyon Coal Company, Storrs, Carbon County, Utah," February 7, 1919, Records of the Federal Mediation and Conciliation Service, Record Group 380 (33–2496) and (33–2991), National Archives.

25. Ibid.

26. John McLennan, telegram to H.L. Kerwin, March 18, 1919, Records of the Federal Mediation and Conciliation Service, Record Group 380 (33–2991), National Archives.

27. Robert G. Bolton and Jasper Metzger to W.N. Wilson, November 18, 1919, Records of the Federal Mediation and Conciliation Service, Record Group 380 (170-923), National Archives.

28. William Bennett to John L. Lewis, December 19, 1919, United Mine Workers of America Records, District 15 Correspondence, 1919.

29. Frank Bonacci manuscript, Utah State Historical Society, Salt Lake City, Utah.

30. E.O. Bill Jackson, interview, Price, Utah, December 11, 1975.

31. *Proceedings of the Twenty-Seventh Consecutive and Fourth Biennial Convention of the United Mine Workers of America, Cleveland, Ohio, September 9, 1919, Volume I, 1919*, Indianapolis, 1919, p. 224.

32. Robert K. Murray, *Red Scare: A Study in National Hysteria, 1919–1920* (Minneapolis, 1955), pp. 153–65.

33. The notes are in the Spring Canyon Area Coal Company Records, MSS 252, Box 3, Brigham Young University Library, Provo, Utah.

34. *Salt Lake Tribune*, October 31, 1919.

35. *Sun* (Price, Ut.), October 31, 1919.

36. *Salt Lake Tribune*, November 1, 1919.

37. Ibid., November 2, 1919.

38. *Salt Lake Herald*, November 1, 1919.

39. *Salt Lake Tribune*, November 12, 1919, and *Deseret News*, November 1, 1919.

40. H.C. Stewart to John McLennan, February 6, 1919, United Mine Workers of America Records, District 15 Correspondence, 1919.

41. Ibid., January 27, 1919.

42. The report was submitted to John L. Lewis and William Green on December 23, 1919, United Mine Workers of America Records, District 15 Correspondence, 1919.

43. H.C. Stewart to John L. Lewis, January 7, 1920, United Mine Workers of America Records, District 15 Correspondence, 1920.

44. John McLennan to John L. Lewis, January 19, 1920, United Mine Workers of America Records, District 15 Correspondence, 1920.

45. O.F. Nigro to John L. Lewis, January 16, 1920, United Mine Workers of America Records, District 15 Correspondence, 1920.

46. The section of the Constitution is quoted in the charges brought against the four men. United Mine Workers of America Records, District 15 Correspondence, 1920.

47. Frank Bonacci to John L. Lewis, February 4, 1920, United Mine Workers of America Records, District 15 Correspondence, 1920.

48. Louis Tikas to John L. Lewis, August 28, 1913, United Mine Workers of America Records, District 15 Correspondence, 1913. See also Zeese Papanikolas, *Buried Unsung: Louis Tikas and the Ludlow Massacre*, pp. 176–77.

49. W.D. Van Horn, A.R. Watkins, J.M. Zimmerman, Committee on Organization to John L. Lewis, March 8, 1920, United Mine Workers of America Records, District 15 Correspondence, 1920.

50. Submitted to John L. Lewis on June 23, 1920, United Mine Workers of America Records, District 15 Correspondence, 1920.

51. Submitted to John L. Lewis on February 3, 1921, United Mine Workers of America Records, District 15 Correspondence, 1921.

52. Minutes of the International Executive Board, August 9–18, 1921, pp. 145–46, United Mine Workers of America Records.

53. The report is found in the United Mine Workers of America Records, District 15 Correspondence, 1920.

Chapter VII

1. *Sun* (Price, Ut.), September 9, 1921.

2. *Wyoming Labor Journal*, August 12, 1921. The officers elected at Kenilworth were president—John Tittle, Sr.; vice president—Walter Brooks; recording secretary—L.R. Evans; financial secretary—Frank Bonacci; trustees—S. Keller and Jas. Dixon; door keeper—A. Johnson.

3. *Sun* (Price, Ut.), March 24, 1922.

4. King, "Statement and Brief," pp. 4–5. This conclusion is supported by the fact that no newspapers reported that Utah had been included in the strike order and the *Wyoming Labor Journal*, semi-official publication for District 22, did not contain any articles on Utah until May 5, more than a month after the strike began.

5. Ibid., p. 8.

6. *Salt Lake Tribune*, April 26, 1922.

7. Ibid., April 1, 1922.

8. *Sun* (Price, Ut.), September 16, 1921.

9. *Salt Lake Telegram*, June 24, 1922.

10. *Salt Lake Tribune*, June 21, 1922.

11. *Salt Lake Telegram*, April 30, 1922.

12. *Wyoming Labor Journal*, May 12, 1922, and August 10, 1923.

13. Helen Zeese Papanikolas, *Toil and Rage in a New Land*, p. 175.

14. *Salt Lake Telegram*, May 1, 1922. A copy of the eviction notice of the Spring Canyon Coal Company is in the Spring Canyon Area Coal Company Records, Box 3, Folder 4, Brigham Young University Library, Provo, Utah. Reports from the Globe Inspection Company's Inspector S-3 are in the same folder. See also interview of George Zoumadakis by Helen Zeese Papanikolas, April 21, 1966, quoted in Helen Zeese Papanikolas, *Toil and Rage in a New Land*, p. 175.

15. Samuel King, "Statement and Brief," pp. 5–6.

16. *Sun* (Price, Ut.), July 21, 1922.

17. *News Advocate* (Price, Ut.), August 3, 1922.

18. Ibid., July 12, 1922.

19. *Wyoming Labor Journal*, July 13, 1923.

20. Henry L. Stimson and Goldthwaithe H. Dorr, "The Campaign of the United Mine Workers of America against the Non-Union Mines of Utah," p. 9. Copy at the Utah State Historical Society. For a discussion of the importance of these organizations to the immigrants, see Notarianni, "Italian Fraternal Organizations in Utah," pp. 172–87.

21. *Deseret News*, April 26, 1922; *Salt Lake Tribune*, April 26, 1922; *Sun* (Price, Ut.), April 28, 1922; and King, "Statement and Brief," pp. 10–11.

22. *Deseret News*, April 28, 1922.

23. *Salt Lake Tribune*, April 28, 29, May 2, 1922; *Salt Lake Telegram*, April 29, 30, 1922; King, "Statement and Brief," pp. 25–26. The newspapers' spelling of "Makesmrticos" is obviously incorrect. The author has not been able to find the true spelling.

24. *Salt Lake Tribune*, April 19, 1922.

25. *Deseret News*, April 29, 1922.

26. *Salt Lake Tribune*, April 28, 1922.

27. *Salt Lake Telegram*, May 1, 1922, and the *Salt Lake Tribune*, May 2, 1922.

28. *Salt Lake Telegram*, May 1, 1922.

29. *Salt Lake Tribune*, May 2, 1922. Apparently some discretion was used in the hiring of the men as deputies. The *Tribune* said that in Salt Lake City only 24 out of 200 applicants were hired.

30. *Wyoming Labor Journal*, June 16, 1922.

31. Ibid.

32. *Salt Lake Telegram*, May 4, 1922.

33. *Wyoming Labor Journal*, June 23, 1922, and July 7, 1922.

34. Ibid., June 23, 1922.

35. For a list of 258 miners from Utah who worked during the strike, see *Wyoming Labor Journal*, September 29, 1922.

36. Helen Zeese Papanikolas, *Toil and Rage in a New Land*, pp. 169–70; *Sun* (Price, Ut.), May 19, 1922, and April 21, 1923; *Salt Lake Tribune*, May 15, 1922; *Salt Lake Telegram*, May 15, 1922; *Eastern Utah Advocate* (Price, Ut.), May 25, 1922; and King, "Statement and Brief," pp. 23–24.

37. *Salt Lake Telegram*, May 18, 1922.

38. *Salt Lake Tribune*, May 17, 1922; *Salt Lake Telegram*, May 18, 1922. The Price newspaper, the *Sun*, deplored Mabey's inconspicuous visit with large headlines reading "Governor Flits In and Out." May 19, 1922.

39. *Salt Lake Tribune*, May 20, 1922.

40. Ibid.

41. Ibid., May 20, 1922, and May 25, 1922.

42. Ibid., May 25, 1922, and *Sun* (Price, Ut.), May 26, 1922.

43. *Salt Lake Tribune*, May 16, 1922, and June 1, 1922.

44. Ibid., June 3, 1922, June 4, 1922, June 10, 1922; *Sun* (Price, Ut.), June 9, 1922; *Deseret News*, June 9, 1922; *Salt Lake Telegram*, June 4, 1922; and King, "Statement and Brief," p. 20. King blamed the company guards as did J. D. Ramsay in a letter dated June 4, 1922, Helper, Utah, and published in the *Wyoming Labor Journal*. A copy of the original letter is in a UMWA scrapbook available at the Helper Public Library, Utah. A microfilm copy of the scrapbook is available at the Utah State Historical Society. Guardsmen having regular employment were allowed to leave the barracks on June 7. Those who were unemployed were held indefinitely in the armories pending further developments in the coal fields. *Salt Lake Telegram*, June 8, 1922.

45. *Sun* (Price, Ut.), June 9, 1922, and the *Salt Lake Telegram*, May 23, 1922.

46. King, "Statement and Brief," p. 20.

47. *Sun* (Price, Ut.), July 14, 1922.

48. The picketing effort was so successful that "if not interfered with would, in all probability, have won the strike for the miners." "Kukis Case—To the Honorable Members of the Pardon Board, State of Utah," George H. Dern Correspondence, January to July 1925, Box 3, State Archives, Salt Lake City, Utah.

49. *Salt Lake Tribune*, June 15, 1922, and *News Advocate*, June 15, 1922.

50. *Salt Lake Telegram*, June 16, 1922. Samuel A. King Papers, Marriott Library, Special Collections, University of Utah, Salt Lake City, Utah.

51. *News Advocate*, June 22, 1922.

52. *State of Utah*, *Biennial Report of the Adjutant General, Utah National Guard, 1921–1922*, pp. 17–18, State Archives, Salt Lake City, Utah. For a detailed discussion of the Guard mobilization and activities in Carbon County during the strike, see Roberts, "History of the Utah National Guard," pp. 219–43. See also *Salt Lake Tribune*, June 15, 1922, and *Deseret News*, June 15, 1922.

53. State of Utah, *Biennial Report of the Adjutant General*, pp. 18–19.

54. Roberts, "Utah National Guard," p. 228.

55. *Salt Lake Tribune*, June 19, 1922, and *Salt Lake Telegram*, June 19, 1922.

56. *Salt Lake Tribune*, June 24, 1922, and *Salt Lake Telegram*, June 24, 1922.

57. *Wyoming Labor Journal*, July 7, 1922.

58. *Deseret News*, July 6, 1922, and July 7, 1922. King, "Statement and Brief," p. 9.

59. King, "Statement and Brief," p. 45.

60. Roberts, "Utah National Guard," pp. 238–39.

61. *Salt Lake Tribune*, September 1, 1922. Two and a half years later, sixty-five Kenilworth miners brought suit against the Independent Coal and Coke Com-

pany to recover wages alleged to have been promised them as a bonus for their loyalty during the 1922 strike. The disillusionment became one of the reasons for a short strike at Kenilworth in 1925. See letter from Frank Bonacci to the *Wyoming Labor Journal*, March 27, 1925, and correspondence regarding the matter in the District 22 Correspondence, 1926, United Mine Workers of America Records, 1926. Frank Bonacci in his letter lists the claims against the coal company for fifteen miners as Chris M. Anderson, $475.00; Roy Thayne, $276.65; Andrian Cracraft, $236.19; Evan J. Griffiths, $312.50; D. J. Griffiths, $491.40; Ernest Anderson, $404.25; George C. Ipson, $312.50; Dean Loveless, $292.50; Vergil Anderson, $387.75; Charles L. Sargent, $536.36; Arnold Anderson, $330.12; Bryce Loveless, $277.10; Leslie Pearson, $396.00; Lorenzo Broderick, $262.50; and Arthur Sargent, $350.92. The bonus was to be at the rate of thirty and a half cents an hour for the number of hours worked during the five-month strike. At Castle Gate there was no problem and all miners who were entitled to the bonus were promptly paid. Tony Priano interview, Helper, Utah, January 7, 1976.

62. *Salt Lake Tribune*, September 6, 1922.

63. Ibid., September 7, 1922, and *Deseret News*, September 8, 1922.

64. King, "Statement and Brief," p. 48.

65. *Wyoming Labor Journal*, September 15, 1922.

66. *Sun* (Price, Ut.), February 16, 1923; King, "Statement and Brief," p. 35; *Wyoming Labor Journal*, March 26, 1926, and June 3, 1926.

67. George Zoumadakis stated that among those arrested for the shooting of Webb "some were chosen who hadn't been there; some who were there were not [arrested]." Helen Zeese Papanikolas, *Toil and Rage in a New Land*, p. 175. Attorney King saw the innocence of the men on trial in the following terms: "There is no question as to the innocence of these men. True, some of the miners fired; true, Webb was killed, but the men who did the firing shot in their own defense. They were not the aggressors. They did not begin the controversy." King to James Morgan, secretary and treasurer, District 22, UMWA, October 16, 1923, in United Mine Workers of America Records, District 22 Correspondence, 1923. See the *Wyoming Labor Journal*, May 11, 1923. The expenses incurred and the need for help from national headquarters are explained in a letter from James Morgan to John L. Lewis, November 2, 1923. See proceedings of the International Executive Board, United Mine Workers of America Records, 1923.

68. Stimson and Dorr, "The Campaign of the UMWA," p. 22.

69. Ibid., p. 23.

70. Utah, Legislature, House, Executive Office, State of Utah, 15th session, 1923, *Journal*, January 10, 1923, p. 27.

71. Mabey failed to win reelection to the governorship. For the reasons for his failure, see Stanford John Layton, "Governor Charles R. Mabey and the Utah Election of 1923" (M.A. thesis, University of Utah, 1969).

Chapter VIII

1. T. L. Burridge Diary, Typescript copy at the Utah State Historical Society, Salt Lake City, Utah.

2. *Salt Lake Tribune*, March 11, 1924, and *Salt Lake Telegram*, March 10, 1924, and March 11, 1924. For a general discussion of the circumstances and causes

of the explosion, see B. W. Dyer, John Crawford, and H. E. Munn, "Report on Castle Gate No. 2 Mine Explosion," Castle Gate, Utah, March 8, 1924, State Archives, Independent Commissions, Castle Gate Relief Committee, 1924–1936, Reports.

3. *Salt Lake Telegram*, March 12, 1924. Other figures were given for the nationality of the dead miners. For example, the Report of Committee on Preliminary Survey of Conditions of Dependency lists American, 74; Greek, 49; Italian, 22; Japanese, 8; English, 7; Austrian, 6; Scotch, 2; Belgian, 1; Negro, 2. State Archives, Independent Commissions, Castle Gate Relief Committee, 1924–1936.

4. *Salt Lake Telegram*, March 10, 1924, and *Salt Lake Tribune*, March 11, 1924.

5. T. L. Burridge Diary, and *Salt Lake Tribune*, March 10, 1924.

6. *Sun* (Price, Ut.), March 14, 1924.

7. *Salt Lake Telegram*, March 10, 1924.

8. *Salt Lake Tribune*, March 10, 1924.

9. *Salt Lake Telegram*, March 12, 1924.

10. Saline Hardee Fraser and Marianne Fraser, "One Long Day That Went On Forever," *Utah Historical Quarterly* 48 (Fall 1980):384, 387–88.

11. *Salt Lake Telegram*, March 15, 1924.

12. Ibid., March 12, 1924.

13. *Sun* (Price, Ut.), April 4, 1924.

14. *Salt Lake Tribune*, March 12, 1924.

15. Ibid., March 14, 1924.

16. *Salt Lake Telegram*, March 14, 1924, and the *Sun* (Price, Ut.), May 16, 1924. Greater light is shed on the living conditions and circumstances after the explosion in a report of a visit by Mrs. Annie Palmer on June 18, 1924: "Woman living in a hovel, with old bench, rickety table boxes and old clothes. Not a decent thing about the place. Had sold chairs for food. Has earned about a dollar and a half since the explosion. A white woman who lives near gives her a loaf of bread each day and a few potatoes, and Edith Hadfield, colored maid at the hotel, gives fifty cents occassionally for food. Woman was not married to Mr. Willis and another woman his wife from whom he has separated is claiming compensation. Cora has lived with him as his wife for seven years. There has never been any question about legality until after explosion." State Archives, Independent Commissions, Castle Gate Relief Committee, 1924–1936, "Cora Willis File."

17. *Salt Lake Tribune*, March 11, 1924.

18. *Sun* (Price, Ut.), March 21, 1924.

19. *Salt Lake Tribune*, March 14, 1924.

20. Fraser and Fraser, "One Long Day That Went On Forever," p. 386.

21. *Sun* (Price, Ut.), March 21, 1924.

22. Ibid., July 31, 1925.

23. Castle Gate Relief Committee, "Georgia Paizakis File."

24. Ibid., Commission Minutes, p. 138.

25. Ibid., "Catherine Pelly File."

26. Ibid., "Myrtle Henderson File."

27. Ibid., "Arley Rollins File."

28. Ibid., Final Report, p. 2.

29. *Sun* (Price, Ut.), July 31, 1925.

30. *Salt Lake Tribune*, March 17, 1924.

31. State of Utah, *Public Documents*, "Report of Industrial Commission," Bulletin No. 4, July 1, 1922–June 30, 1924, pp. 43–47.

32. *Salt Lake Tribune*, February 10, 1930.

33. *Sun* (Price, Ut.), July 5, 1927.

Chapter IX

1. *Wyoming Labor Journal*, April 20, 1923.

2. Ibid., October 26, 1923.

3. Frank Bonacci manuscript, p. 24, and the *Sun* (Price, Ut.), July 19, 1924.

4. *Wyoming Labor Journal*, April 19, 1924.

5. Arthur McArthur to Thos. R. Stockett, August 12, 1924, Spring Canyon Area Coal Company Records, MSS 252, Box 3, Folder 4, Brigham Young University Library, Provo, Utah.

6. "Extracts from Reports of Inspector S-2," Sunnyside, August 21, 1924, Spring Canyon Area Coal Company Records, MSS 252, Box 3, Folder 4, Brigham Young University Library, Provo, Utah.

7. "Extracts from Report of Inspector S - - - - -," Castle Gate, February 18, 1925, Spring Canyon Area Coal Company Records, MSS 252, Box 3, Folder 4, Brigham Young University, Library, Provo, Utah.

8. Ibid., February 23, 1925.

9. Ibid., February 18, 1925.

10. Ibid., February 23, 1925.

11. *Deseret News*, April 17, 1925, and *Salt Lake Telegram*, April 17, 1925.

12. Samuel King to James Morgan, May 9, 1925, Samuel King Papers, Marriott Library, University of Utah, Salt Lake City, Utah.

13. *Salt Lake Telegram*, April 19, 1925.

14. *Sun* (Price, Ut.), June 5, 1925.

15. "Extracts from Report of Inspector S-2," Sunnyside, March 20, 1926, Spring Canyon Area Coal Company Records, MSS 252, Box 3, Folder 4, Brigham Young University Library, Provo, Utah.

16. "Extracts from Report of S - - - -," Helper, January 9, 1926, Spring Canyon Area Coal Company Records, MSS 252, Box 3, Folder 4, Brigham Young University Library, Provo, Utah.

17. Ibid., March 27, 1926.

18. "Extracts from Report of Inspector S-3," March 24, 1926, Spring Canyon Area Coal Company Records, MSS 252, Box 3, Folder 4, Brigham Young University Library, Provo, Utah.

19. "Extracts from Report of Inspector S-2," April 23, 1926, Spring Canyon Area Coal Company Records, MSS 252, Box 3, Folder 4, Brigham Young University Library, Provo, Utah.

20. "Extracts from Report of Inspector S-3," June 12, 1926, Spring Canyon Area Coal Company Records, MSS 252, Box 3, Folder 4, Brigham Young University Library, Provo, Utah.

21. "Extracts from Report of Inspector S-2," March 20, 1926, Spring Canyon Area Coal Company Records, MSS 252, Box 3, Folder 4, Brigham Young University Library, Provo, Utah.

22. Spring Canyon Coal Company Annual Report, 1926–1927, p. 46, Spring Canyon Area Coal Company Records, MSS 252, Box 3, Folder 4, Brigham Young University Library, Provo, Utah.

23. "Extracts from Report of Inspector S-3," April 19, 1927, Spring Canyon Area Coal Company Records, MSS 252, Box 3, Folder 4, Brigham Young University Library, Provo, Utah.

24. *Rock Springs Miner* (Rock Springs, Wy.), March 18, 1927.

25. *Sun* (Price, Ut.), April 6, 1928.

26. James Morgan to John L. Lewis, March 26, 1929, and April 10, 1929, United Mine Workers of America Records, District 22 Correspondence, 1929. In his manuscript Frank Bonacci reported that he "resigned the union job." Given Bonacci's loyalty and commitment to the UMWA, it seems more likely that he was forced out of his job rather than resigning voluntarily.

27. Morgan to Lewis, March 26, 1929, United Mine Workers of America Records, District 22 Correspondence, 1929.

28. Frank Bonacci Record Book, p. 24, Utah State Historical Society.

29. Lewis to Morgan, April 22, 1929, United Mine Workers of America Records, District 22 Correspondence, 1929.

30. Martin Cahill and George Young to John L. Lewis, May 13, 1929, United Mine Workers of America Records, District 22 Correspondence, 1929.

31. John L. Lewis to Martin Cahill, May 21, 1929, United Mine Workers of America Records, District 22 Correspondence, 1929.

32. Proceedings of the International Executive Board, United Mine Workers of America, p. 289, September 18, 1929.

33. Spring Canyon Coal Company Annual Report, 1927–1928, pp. 37–38, Spring Canyon Area Coal Company Records, Brigham Young University, Provo, Utah.

34. *Deseret News*, October 3, 1931, *Salt Lake Tribune*, October 3, 1931, and the *Sun* (Price, Ut.), October 8, 1931.

35. Mrs. J. E. Tadlock to Governor George H. Dern, January 6, 1930, George Dern Correspondence, Box 39, Folder T, State Archives, Utah.

36. Steve Zamof to George H. Dern, July 17, 1930, George Dern Correspondence, Box 40, Folder U-Z, State Archives, Utah.

37. Hugh McMillian to Henry H. Blood, May 23, 1933, Henry H. Blood Correspondence, Box 7, Folder Mc, State Archives, Salt Lake City, Utah.

Chapter X

1. For example, Arthur M. Schlesinger, Jr., writes: "No one understood better that labor's position under the Act depended on its power—and thus on its success in exploiting the new government-sponsored organizing possibilities—than John L. Lewis. . . . Few other labor chiefs saw the opportunity with Lewis's clarity. Most old-line AF of L leaders, reared in the voluntaristic tradition of Samuel Gompers, looked on government with mistrust." Arthur M. Schlesinger, Jr., *The Age of*

Roosevelt: The Coming of the New Deal (Boston, 1958), pp. 137, 140. See also Irving Bernstein, *The Turbulent Years: A History of the American Worker 1933–1941* (Boston, 1969), pp. 40–41. Boasting of the UMWA accomplishment at the 1934 UMWA International Convention Lewis offered the same conclusion. "Truly a great accomplishment; Truly Marvelous progress! Greater progress, may I say with pardonable pride, than has been made by any other trade union organization in America. . . . The fact that the United Mine Workers of America has made more progress under the policies of the Industrial Recovery Act than has, perhaps been the case with other basic industries is merely indicative of the fact that, having promoted the legislation, we, perhaps, understood its potentialities and its possibilities to a greater degree than others; we, perhaps, worked longer days and longer nights in carrying the message to our people and encouraging them to join in the great undertaking." Quoted in George Korson, "A History of the UMWA, Chapter 10," in *United Mine Workers Journal*, February 1, 1966.

2. Quoted in Arthur M. Schlesinger, Jr., *The Age of Roosevelt: The Coming of the New Deal*, p. 138–39.

3. Quoted in George Korson, "A History of the UMWA, Chapter 10," p. 6.

4. Helen Zeese Papanikolas, "Unionism, Communism, and the Great Depression: The Carbon County Coal Strike of 1933," *Utah Historical Quarterly* 41 (Summer 1973) :260.

5. Telegram sent by the Standard Coal Miners to Governor Blood, May 8, 1933, in Henry H. Blood Correspondence, Box 2, Folder Carbon County Riot, State Archives.

6. Telegram, George Cole to John L. Lewis, May 9, 1933, in United Mine Workers of America Records, District 22 Correspondence, 1933.

7. Telegram John L. Lewis to George Cole, May 10, 1933, in United Mine Workers of America Records, District 22 Correspondence, 1933.

8. George Cole to John L. Lewis, May 11, 1933, United Mine Workers of America Records, District 22 Correspondence, 1933.

9. John L. Lewis to George Cole, May 16, 1933, United Mine Workers of America Records, District 22 Correspondence, 1933.

10. John L. Lewis to George Cole and Tony Radalj, May 27, 1933, United Mine Workers of America Records, District 22 Correspondence, 1933. John L. Lewis had little regard for the organizing abilities of Radalj as noted earlier.

11. Tony Radalj to John L. Lewis, June 3, 1933, United Mine Workers of America Records, District 22 Correspondence, 1933.

12. See Irving Bernstein, *Turbulent Years: A History of the American Worker, 1933-1941*, pp. 40–41.

13. Charles Guynn, "National Miners Union Built Through Struggle," *Carbon County Miner*, June 21, 1933, June 28, 1933, and July 22, 1933.

14. *National Miners Union Membership Book* (Women's Auxiliary, Book No. 2724), South Slavic Archives, Special Collections, Marriott Library, University of Utah, Salt Lake City, Utah.

15. Regarding reports on the National Miners Union meeting, Helen Zeese Papanikolas notes that National Guard reports on the National Miners Union meetings for May 28, June 4, 11, 18, 25, August 6, 13, 1933, and other material pertaining to the strike were salvaged from National Guard files by a son-in-law of Varro C. Jones, secretary of the UMWA in Kenilworth during the strike. Photocopies, donated

by Rolla West, have been placed in the Special Collections Section, Marriott Library, University of Utah. National Guard reports on NMU meetings of August 31 and September 1, 1933, are included in Henry H. Blood Correspondence, 1933, Box 2, Folder Carbon County Riot, State Archives. Two of the reports are signed "Cavenah" and "Reese." Helen Zeese Papanikolas, "The Carbon County Coal Strike of 1933," p. 261. Hereafter these reports will be cited as NMU meeting reports with date.

16. Helen Zeese Papanikolas, "The Carbon County Coal Strike of 1933," p. 264.

17. NMU Meeting Report, May 28, 1933, and the *Carbon County Miner*, June 7, 1933.

18. The earliest issue of the *Carbon County Miner* available is June 7, 1933. This issue is identified as Volume 1, Number 2.

19. Helen Zeese Papanikolas, "The Carbon County Coal Strike of 1933," p. 260.

20. *Carbon County Miner*, June 14, 1933.

21. Ibid., June 21, 1933.

22. Nicholas Fontecchio to Philip Murray, June 15, 1933, United Mine Workers of America Records, District 22 Correspondence, 1933.

23. The Utah unions were listed in this order in a *Wyoming Labor Journal* article, September 15, 1933. Although the numbers are not in sequence, this is perhaps the order in which the local union charters were applied for rather than received. For example, Wattis miners took great pride in the fact that their local was the first organized in 1933 and they applied for a local charter several days in advance of Standardville. However, the Standardville local was given the lowest number in Utah, 5806, suggesting that it had been the first organized. Leland Powell interview.

24. *Carbon County Miner*, June 21, 1933, June 28, 1933, and July 14, 1933. Also reflective of this attitude was a letter written to the editor of the *Carbon County Miner* by a miner from Peerless and published in the June 28, 1933, edition.

"I find that some of the coal companies are encouraging and trying to get their employees to join the United Mine Workers of America. Now, fellow miners of Carbon County, have you forgotten what the UMWA did to you in 1922? Have you forgotten the train load of scabs that was held up at the mouth of the Utah railroad tunnel near Helper?

Now fellow workmen, is it possible that after the Standard Coal Co. after fighting you and throwing you out of your homes, getting the state militia to hold you out of the canyon, is now asking you to join the UMWA? I want to warn you men to watch your step. The UMWA organizers have had their chance. They have failed. You know and I know when the Standard Coal Co. and UMWA join hands and coax you to join them someone is going to be the loser. And, my fellow workmen, YOU are going to be the loser. Now you miners and other workmen of Standardville have a chance for a new deal.

Join the National Miners Union. Have a Union of your own; join a Union that will stand back of you when you want to better your working conditions; join a Union in which the superintendent will not be at the head."

25. Telegram from Nicholas Fontecchio to Governor Henry H. Blood, August 25, 1933, Henry H. Blood Correspondence, Box 2, Folder Carbon County Riot, State Archives; and Frank Bonacci to John L. Lewis, August 7, 1933, United Mine Workers of America Records, District 22 Correspondence, 1933.

26. Nicholas Fontecchio, "The New Deal For the Coal Mining Industry," United Mine Workers of America Records, District 22 Correspondence.

27. A resolution adopted by the United Mine Workers of America representatives in a meeting held at the Price American Legion Hall, August 21, 1933, was sent to Governor Henry H. Blood and the law enforcement officers of Carbon County. The resolution declared:

"WHEREAS, the agents of the Communist Party have been for some time in our county recruiting people for their party, largely irresponsible, of a bad repute, and enemies of our government; and

WHEREAS, this element has been continuously running at large abusing the good people of our county and have now resorted to the most vicious means of terrorizing and interfering with the lives, liberty and the pursuit of happiness of our people.

Resolved, that we, the representatives of 85% of the coal miners employed in the State of Utah, loyal American citizens and members of the United Mine Workers of America, emphatically resent the presence and the conduct of these enemies of our people and our government in our county and we demand that the law enforcement officers of the state and county will at once take steps to penalize the Communist leaders and their aids to the extent of removing them from our midst; we further

RESOLVE, that we, home-loving, law-abiding citizens, can not tolerate the present existing situation much longer; and we so advise the authorities of the state and county to act in this emergency before something serious may take place."

The resolution is in the Henry H. Blood Correspondence, Box 11, Utah Federation of Labor Folder, State Archives.

28. *Carbon County Miner*, June 14, 1933.

29. Ibid., July 15, 1933.

30. NMU Meeting Report, September 4, 1933.

31. Copy of flyer in author's possession.

32. *Carbon County Miner*, July 22, 1933.

33. Helen Zeese Papanikolas, "The Carbon County Strike of 1933," p. 260.

34. NMU Meeting Report, May 28, 1933. In the late 1940s and early 1950s, Paul Crouch became an important witness in the Communist witch hunts of the McCarthy Era. He testified that he had been a member of the Communist party from 1925 to 1942. According to Roy M. Cohn, Crouch was responsible for furnishing Joe McCarthy with the "information" that led to the Fort Monmouth hearings. Willard Shelton, "Paul Crouch, Informer," *The New Republic*, July 19, 1954, pp. 7–9. For a report of Crouch's testimony before the National Congress, see U.S. Congress, House of Representatives, Eighty-First Congress, First Session, Committee on Un-American Activities, "Hearings Regarding Clarence Hiskey including testimony of Paul Crouch." The testimony of Paul Crouch was given to the House Un-American Activities Committee on May 24, 1949.

35. NMU Meeting Report, August 6, 1933.

36. NMU Meeting Report, May 28, 1933.

37. *Helper Journal*, August 18, 1933.

38. *Carbon County Miner*, August 19, 1933.

39. *Helper Journal*, July 7, 1933. See Rolla E. West, typewritten manuscript on the Carbon County Strike of 1933, pp. 47–51, Utah Historical Society, Salt Lake

City, Utah, for a description of the decision to deny the NMU permission to participate in the parade.

40. This statement is attributed to Margaret Nemanich. See Helen Zeese Papanikolas, "The Carbon County Strike of 1933," p. 272. See also *Helper Journal*, July 7, 1933.

41. *Salt Lake Tribune*, July 5, 1933.

42. *Sun-Advocate* (Price, Ut.), July 6, 1933.

43. *Salt Lake Tribune*, July 11, 1933.

44. *Helper Journal*, July 7, 1933.

45. NMU Meeting Report, July 9, 1933.

46. *Carbon County Miner*, August 19, 1933.

47. *Helper Journal*, August 11, 1933.

48. *Carbon County Miner*, August 19, 1933.

49. *Sun-Advocate* (Price, Ut.), August 24, 1933.

50. *Carbon County Miner*, August 19, 1933.

51. *Sun-Advocate* (Price, Ut.), August 24, 1933.

52. *Carbon County Miner*, September 2, 1933.

53. Telegram, Nicholas Fontecchio to Governor Henry H. Blood, August 25, 1933. Henry H. Blood Correspondence, Box 2, 1933, Folder Carbon County Riot, State Archives.

54. *Sun-Advocate* (Price, Ut.), October 19, 1933.

55. Rolla West manuscript, pp. 57–59.

56. *Helper Journal*, August 25, 1933.

57. Frank Bonacci to John L. Lewis, January 6, 1934, United Mine Workers of America Records, District 22 Correspondence, 1934.

58. Rolla West was strongly criticized by some local residents especially from the Helper area for the leadership he assumed in organizing and leading the deputies. Local residents were unaware of the arrangement giving West special powers as the chief deputy sheriff of Carbon County, and were convinced that West was usurping authority far beyond his jurisdiction as Price City mayor. No contemporary account of the arrangement has been found and the manner in which West was asked to assume control of the situation is somewhat suspect. According to West he was not contacted directly by Governor Blood, but by Preston G. Peterson, a member of the State Road Commission. In his manuscript West writes: "Preston G. Peterson, member of the state road Commission presented himself at Bill Reid's service station in Price and asked for an audience with Bill Reid and Rolla West. Rolla West was sent for, and when he met with Press Peterson and Bill Reid, Mr. Peterson informed them that he had been asked by Governor Henry H. Blood to be his personal representative in Carbon County to achieve a rational solution of the problems in Carbon County and that the Governor had requested him to ask Bill Reid to be his aide and contact with the Carbon County Government. And the Governor specifically asked that Rolla West accept a position of Chief Deputy Carbon County Sheriff, to organize and direct a force of Carbon County men as deputy sheriff to resolve the troubles in the county. Rolla West was to have a free hand and be responsible only to the governor of the state. Bill Reid may have known something of this planning but Rolla West certainly was taken by surprise. . . . West's first thoughts were to say 'no' but he knew that he could not, in good grace, say 'no.' He asked if the Sheriff

knew of this plan, the reply was that he did not as yet know of these arrangements. . . . West asked for a conference with the Sheriff and the Sheriff was sent for and came down. The arrangement was explained to him, and West informed the sheriff that he had no intention of embarrassing the sheriffs office, that he had no political ambitions and would plan to take no action secret from the sheriff, and that he was accepting this assignment only with the hope of helping to bring order to the situation in Carbon County." Rolla West manuscript, pp. 60–62.

59. Ibid., p. 63.

60. *Deseret News*, August 25, 1933.

61. Rolla West manuscript, p. 73.

62. Ibid., p. 74.

63. Ibid., p. 83.

64. Ibid., p. 84.

65. Ibid., p. 86.

66. *Helper Journal*, September 1, 1933.

67. Rolla West manuscript, pp. 89–91. West described the incident as follows: "He heard a lady's angry voice coming from up the railroad tracks. She was very much annoyed about something expressing herself in unladylike terms for all to hear. She was looking for that 'so and so Mister, a Mayor a West.' As she came down to where West was standing by the road which paralleled the R.R. Track at that place she espied West and crossed over to him swearing profusely—'you so and so and so you hold a my husbon prisoner. He gotta coma home. Fixa the bread for the customer.' And she stepped up to West and started flailing him in the face with her hand satchel, bloodying his nose. He grabbed her hands to keep her from flailing him with her hand satchel and she kicked him on the shins, so he held her at arms length away from him. 'Who is this woman?' he asked. Remembering Dave Parmley from up at Consumers. West didn't know what to expect next. The men seemed to get a great kick out of all this. Finally two deputies came up and told him she was the baker's wife and said they would take care of her. 'Put her over in the Ford with the old man,' West said and proceeded to clean up his bloodied face and rub his shins. He turned around to speak to the driver of a car that drove up and he heard someone say, 'Slam the door on her g d legs if she won't put them in the car.' It was the clerk of the mine office, who had no connection with the deputies at all. The deputies were trying to put the irate lady in the back seat of the car with her husband and he was saying 'Now, mama, get in the car. Let us not have trouble.' But Mamma wasn't getting into the car even with the insistent assistance of two deputies. The mine clerk started for the car door and West hurried over there but the mine clerk beat him there and had slammed the door on the lady's legs before West reached him. West pushed him away and told him to get back up in the mine office and stay there and mind his own business and the deputies would take care of theirs." West was later sued by the baker's wife for $20,000, however, the jury's decision went against the woman.

68. Rolla West manuscript, p. 92.

69. Ibid., p. 98.

70. Ibid., pp. 98–101.

71. *Helper Journal*, September 1, 1933.

72. Rolla West manuscript, p. 117.

73. Ibid., p. 122.

74. Helen Zeese Papanikolas, "The Carbon County Coal Strike of 1933," p. 282.

75. *Sun-Advocate* (Price, Ut.), August 31, 1933.

76. *Carbon County Miner*, September 2, 1933. In addition, see a broadside of the same nature as the newspaper article, in Henry H. Blood Correspondence, 1933, Box 2, Folder Carbon County Riot, State Archives.

77. NMU Meeting Report, September 4, 1933.

78. *Sun-Advocate* (Price, Ut.), September 7, 1933.

79. Rolla West manuscript, p. 137.

80. Helen Zeese Papanikolas, "The Carbon County Coal Strike of 1933," pp. 284–85.

81. *Helper Journal*, September 15, 1933.

82. *Sun-Advocate* (Price, Ut.), September 14, 1933.

83. "Stenographic Report of Meeting of a Protest Body on the Carbon County Labor Problem at the First Congregational Church Salt Lake City, Utah, September 15, 1933, 7:30 p.m."

84. *Helper Journal*, September 15, 1933.

85. "Stenographic Report," p. 30.

86. *Progressive Independent* (Salt Lake City), September 22, 1933. This issue of the *Progressive Independent* contains resolutions adopted and statements by members of the investigating committee.

87. The telegrams, letters, and resolutions are in the Henry H. Blood Correspondence, 1933, Box 2, Folder Carbon County Riot, State Archives.

Many of the resolutions, especially those from outside Utah, did not accurately describe the Utah situation. However they do reflect a strong sense of indignation with conditions. The following letter of protest from Charlotte, North Carolina, is typical of the nearly fifty letters, telegrams, and resolutions sent to Governor Blood during the first two weeks of September.

"We the workers of Charlotte N.C. assembled in a mass meeting on Sept. 4th, strongly protest against the cold blooded brutal murder of the striking Utah coal miners, whose only 'crime' is that they are demanding bread for themselves and their children. As a pack of blood-hounds your troops are preying down upon the innocent and defenseless miners and their children, killing and terrorising them by the scores. All this is done in the quest for huge profits under your sanction and orders.
We hold you responsible for the lives of these miners and must energetically demand the immediate removal of all troops from the strike area and the immediate and unconditional release of all miners and organisers that have been arrested and held for no other crime than organising workers."

88. *Progressive Independent* (Salt Lake City), September 22, 1933.

89. Robert K. Murray, *Red Scare: A Study in National Hysteria 1919–1970* (Minneapolis, 1955), p. 233.

90. *Carbon County Miner*, October 7, 1933.

91. Copy available at Utah State Historical Society, Salt Lake City, Utah.

92. *Pacific Reporter*, State V. Guynn, NO. 5597, Supreme Court of Utah, September 7, 1935. For an account of the strategy used in winning the new trial, see Harry Metos interview by Helen Zeese Papanikolas and Ted Paulos, December 7, 1971, transcript at the Western History Center, University of Utah, Salt Lake City, Utah.

93. *Carbon County Miner*, October 28, 1933.

94. *Sun-Advocate* (Price, Ut.), April 19, 1934.

95. Ibid., May 3, 1934.

96. Charles Grant, Secretary Workers and Farmers Protective Union to Governor Blood, June 28, 1933, Henry H. Blood Correspondence, 1933, Box 2, Folder Carbon County Riot, State Archives.

97. *Helper Journal*, September 29, 1933.

98. Frank Bonacci to John L. Lewis, December 6, 1934, United Mine Workers of America Records, District 22 Correspondence, 1935.

99. John L. Lewis to Frank Bonnaci, December 11, 1934, United Mine Workers of America Records, District 22 Correspondence, 1935.

100. Rolla West manuscript, pp. 143–44.

101. Ibid., p. 53.

102. Nicholas Fontecchio to Philip Murray, September 7, 1933, United Mine Workers of America Records, District 22 Correspondence, 1933.

103. A. T. Fotland to William Knerr, July 21, 1933, Records of the Federal Mediation and Conciliation Service, Record Group 380 (176-61), National Archives.

104. George Lambert to John L. Lewis, February 11, 1934, United Mine Workers of America Records, District 22 Correspondence, 1934, File Columbia Steel Company, Utah.

105. *Sun-Advocate* (Price, Ut.), September 21, 1933.

106. *Wyoming Labor Journal*, December 1, 1933.

107. Philip Murray to Nicholas Fontecchio, November 20, 1933, United Mine Workers of America Records, District 22 Correspondence, 1933.

108. Irving Bernstein, *Turbulent Years: A History of the American Worker, 1933-1941*, p. 46.

109. Ibid., pp. 41–52, 56–60.

110. Orvel Peterson to John L. Lewis, August 18, 1933, United Mine Workers of America Records, District 22 Correspondence, 1933.

111. James Morgan to John L. Lewis, November 21, 1933, United Mine Workers of America Records, District 22 Correspondence, 1933, File Columbia Steel Company, Utah.

112. George Lambert to John L. Lewis, February 11, 1934, United Mine Workers of America Records, District 22 Correspondence, 1934.

113. John L. Lewis to George Lambert, January 18, 1934, United Mine Workers of America Records, District 22 Correspondence, 1934.

114. John L. Lewis to Frank Bonacci, February 24, 1934, United Mine Workers of America Records, District 22 Correspondence, 1934.

115. George Creel to Allen T. Sanford, February 24, 1934, Henry H. Blood Correspondence, 1933, Box 2, Folder Carbon County Riot, State Archives.

116. *Sun-Advocate* (Price, Ut.), March 1, 1934.

Chapter XI

1. Rex Lauck, ed., *John L. Lewis and the International Union United Mine Workers of America* (Washington, D.C., 1952), p. 65.

2. *Sun-Advocate* (Price, Ut.), June 4, 1936.

3. Frank Bonacci, "Record Book." Copy obtained from his daughter, Mrs. Marian Bonacci Lupo, and filed with the Utah State Historical Society.

4. Ernest W. Randa and Reed C. Richardson, *Union Membership in Utah*, Bulletin No. 1, Institute of Industrial Relations, University of Utah, 1960.

5. *Sun-Advocate* (Price, Ut.), November 4, 1943.

6. Ibid., June 22, 1943.

7. Ibid., March 8, 1945.

8. Reynolds, comp., *Centennial Echoes From Carbon County*, pp. 145–46.

Chapter XII

1. *Official Proceedings of the Sixteenth Annual Convention, Western Federation of Miners, July 13th to July 29th, 1908*, Denver, Colorado, 1908, p. 291.

2. David Brody, *Steelworkers in America: The Non-union Era* (Cambridge, Mass., 1960), pp. 86–87.

3. Martin Cahill and George Young to John L. Lewis, May 13, 1929, United Mine Workers of America Records, District 22 Correspondence, 1929.

4. Quoted in Victor R. Green, *The Slavic Community on Strike: Immigrant Labor in Pennsylvania Anthracite* (Notre Dame, Ind., 1968), p. 208.

5. Gerald Rosenblum, *Immigrant Workers: Their Impact on American Labor Radicalism* (New York, 1973).

6. Robert H. Zieger's review of Gerald Rosenblum's *Immigrant Workers: Their Impact on American Labor Radicalism*, in *Labor History* 15 (Winter 1974): 111.

7. David Brody, *Steelworkers in America: The Non-union Era*, pp. 136–37.

8. Ibid., p. 140.

9. Helen Zeese Papanikolas, "Greek Towns," *Toil and Rage in a New Land*, p. 143.

10. Davies, "LDS Labor Philosophy," p. 389.

11. John McLennan to William Green, October 28, 1918, United Mine Workers of America Records, International Executive Board Documents and Circulars Files.

12. John J. Battagigliotti interview, Ogden, Utah, February 20, 1976.

13. *United Mine Workers Journal*, December 10, 1903.

Index